FEARLESS CHANGE AND SOCIAL ACTION IN DIFFICULT TIMES

Our divided politics, unable to solve the challenges we face concerning society's hierarchies of injustice, poverty, endless war, and climate change, are now backtracking to even more division. But the reality goes far deeper than the simple politics of left and right. For true change, we need something more profound: a culture shift, a collective change of consciousness.

Fearless Change and Social Action in Difficult Times argues that culture shifts don't just happen, they require a strong focus on social and cultural human connection which neither political nor economic power can provide alone. It is only deep participation and social integrative power which have the capacity to create these necessary cultural and societal transformations. Developing awareness in participatory groups of thought-worlds which remain out-of-sight but give cover to the implicit rules of culture and society is the first step to creating shared awareness of constructs and negative thought-worlds that subconsciously support inequality. Consciously putting aside those that are negative allows for the emergence of new positive realities and social movements. Thus, the real revolution is of the mind. It does take courage, but this is the process by which better futures are created.

Offering significant contributions to sociology and social theory, this book promotes an understanding that societal change is rooted in social power and cultural shifts. Inclusive in its presentation, students, professors, NGO professionals, volunteers, activists, and interested observers will find this book of high interest.

Paula Donnelly Roark is a social science researcher, writer, and activist. She previously served as Senior Social Scientist at the World Bank, Director of Research and Evaluation at the African Development Foundation, and as an advisor to the United Nations' UNICEF, UNDP, and British government aid. She is the author of *Social Justice and Deep Participation: Theory and Practice for the 21st Century* (2015).

"This book addresses current societal challenges such as injustice, poverty, war, and climate change. The conceptual and methodological issues that form the basis for this book are centered around the idea that societal challenges require a culture shift and a collective change of consciousness. The author argues that developing awareness in participatory groups is key to creating new positive realities and social movements. In this, it offers a unique and significant contribution no other text currently on the market cover. It will be a must-have for students, activists, and organizers who want to make real and lasting change."
       Christopher Pieper, *Senior Lecturer of Sociology, Baylor University*

"This book offers a vital contribution to the subject of sociology by promoting an understanding that societal change is rooted in cultural and consciousness shifts. The author's approach to examining the depth of societal issues through the lens of social and cultural human connections proposes an innovative pathway to address contemporary challenges. This can be especially significant for students, who are often at the forefront of social movements and are looking for more profound theoretical and practical frameworks to make sense of and impact their world."
       James Stobaugh, *Associate Professor of Sociology,*
       Arkansas Tech University

"We need this book more than ever! In a time of intense political and cultural polarization, Donnelly Roark enters the fray not as another contentious voice but as a cultural healer. Confronting directly what she calls the "four ancient wrongs" of injustice, endless war, poverty, and climate crisis, she offers a truly novel analysis of both the root causes and the solutions to these wrongs. The analysis turns traditional sociological analysis on its head and starts with belief systems, cultural change, social solidarity, mutual trust, and deep participation as keys to institutional change."
       Stephen Valocchi, *Professor of Sociology, Trinity College, author of*
       Social Movements and Activism in the USA *(Routledge)*

"A lively book that analyses the intersections of injustice and offers tools for transformation. Based on a lifetime of international experience working for social justice, Donnelly Roark proposes a way forward."
       Lesley J. Wood, *Associate Professor of Sociology, York University*

FEARLESS CHANGE AND SOCIAL ACTION IN DIFFICULT TIMES

Exploring Sociological Insights for Social Transformation

Paula Donnelly Roark

Routledge
Taylor & Francis Group
NEW YORK AND LONDON

Designed cover image: Chris Stenger, Unsplash

First published 2025
by Routledge
605 Third Avenue, New York, NY 10158

and by Routledge
4 Park Square, Milton Park, Abingdon, Oxon, OX14 4RN

Routledge is an imprint of the Taylor & Francis Group, an informa business

© 2025 Paula Donnelly Roark

The right of Paula Donnelly Roark to be identified as author of this work has been asserted in accordance with sections 77 and 78 of the Copyright, Designs and Patents Act 1988.

All rights reserved. No part of this book may be reprinted or reproduced or utilised in any form or by any electronic, mechanical, or other means, now known or hereafter invented, including photocopying and recording, or in any information storage or retrieval system, without permission in writing from the publishers.

Trademark notice: Product or corporate names may be trademarks or registered trademarks, and are used only for identification and explanation without intent to infringe.

Library of Congress Cataloging-in-Publication Data
Names: Donnelly-Roark, Paula. author.
Title: Fearless change and social action in difficult times : exploring sociological insights for social transformation / Paula Donnelly Roark.
Description: New York, NY : Routledge, 2024. | Includes bibliographical references and index.
Identifiers: LCCN 2024023993 | ISBN 9781032789248 (hardback) | ISBN 9781032778778 (paperback) | ISBN 9781003489771 (ebook)
Subjects: LCSH: Social change. | Social action. | Sociology--Philosophy.
Classification: LCC HM831 .D647 2024 | DDC 303.4--dc23/eng/20240611
LC record available at https://lccn.loc.gov/2024023993

ISBN: 978-1-032-78924-8 (hbk)
ISBN: 978-1-032-77877-8 (pbk)
ISBN: 978-1-003-48977-1 (ebk)

DOI: 10.4324/9781003489771

Typeset in Sabon LT Pro
by KnowledgeWorks Global Ltd.

*For Cheyenne, Collin, and Tatiana
part of the next generation and its enduring promise*

CONTENTS

Acknowledgements   xi

1 Introduction   1

Changing Perspectives 1
The Four Ancient Wrongs 4
'Why' and 'How' Are the Questions 6
Culture Shifts and Thought-Worlds 8
Creating Fearless Change 9
Departure Points 12

PART I
Investigating the Four Ancient Wrongs   15

2 Injustice: The Malice of Racism   17

Those of Us Who Think We Are White 19
Racism Up Close 22
Crimes of Being 26
Justice Ignored 27
Branded as 'Less Than' 34

3 Injustice: The Oppression of Gender   39

Masculine Systems 39
Tinkering Around the Edges 43

Hidden Histories 48
Creating Collective Identities 51
Alternative Futures 53
Yearning 54

4 Injustice: The Invisibility of Class ... 58

Explorations 59
Forgotten History 59
Race-Talk Heritage 64
The Not-So-Magical Whitening of America 69
Voting and Slumming 71
Bamboozled 74

5 Endless War ... 77

Serving American Interests 78
Accountability Resistance 80
Eroding the Peaceable Bias 83
Playing Outlaw 87
The War at Home 92
Sanctification: Guns, War, and Violence 96

6 Suffering of Poverty ... 102

Rich-Country Poverty: Facts and Status 102
Consequences 104
Political Undertow and Its Mythologies 107
Debates Behind the Strategies 110
The Devastated and Exploited 113
Rejecting Indifference 118

7 Planet Plunder ... 122

A Crisis of Rising Temperatures 123
Carbon Corruption 125
Delay and Denial 129
New Global Frameworks 133
The Inequalities of Inequities 135
Mending Hope and Panic Together 138

PART II
Activating Social Power — **145**

8 Why Social Power? — 147

Investigation Results 148
Inter-Twining of Inequality 151
Current Social Thought 154
From Demise to Genesis 157
The Power of the Social 161
Questions 164

9 What We Say Is Not What We Do — 168

Excluders, Includers, and Fence-Straddlers 169
The Paradox Syndrome 171
Turn-About Experts 174
Earnestly Fooling Ourselves 177
Self-Induced Confusion 179
Thinking Differently 181

10 Our Worldview No Longer Fits Reality — 183

First Revolutions 183
Divisions of Right and Left 185
Immorality and Compromise 187
Return of the Shock Troops 189
None of Us Understood 191
The Root Cause 194

11 Solving the Unsolvables — 197

Crumbling Justifications 197
Disabling Thought-World Camouflage 200
The Negative in the Positive 204
Enduring Complicities 206
Moving Beyond Politics 208
Reset for Reality 210

PART III
Culture Shift Action — 215

12 Exiting the Circle of Indifference — 217

Below the Decibel of Politics 217
Changing World Views 219
Complexity and the Relational Mind 221
Constructing Mutual Genealogies 224
Practicing Equality and Liberty Together 227
There Are Enough of Us 230

13 Deep Participation Practice — 233

Retrieving the Good and the True 234
From Practice to Theory 238
Real World Starting Points 242
Deep Participation's Six Elements 245
New Game, New Rules 251
Courage in Solidarity 253

14 Organizing for Culture Shifts — 255

Traditional Social Movements 256
Deep Participation Social Movements 258
Find Your Group 260
The Three R's 261
Umbrella Leverage 266
Culture Shift Resources 268

Index — 273

ACKNOWLEDGEMENTS

I always remain indebted to a group of friends and colleagues who are scholars, participation practitioners, and social change agents, with whom I have worked over the years on the continent of Africa, including my long-term friend Grace Hemmings, whose work spans both continents. But this book, because of the circumstances of time and subject, has primarily relied on a relatively small but wonderful circle of friends and family of diverse expertise who have supported and sustained me over the past six years.

The early shaping of this book took some work to adequately focus on its diverse subjects. Fortunately, it was strongly assisted by two groups, which happily occasionally intermixed. The first was a circle of friends meeting together every week, starting in 2018, named "wine and whine." The second was weekly Zoom meetings, particularly as Covid hit, with extended family members spread around the globe, which also became a not-to-miss weekly get-together.

As a result of these diverse discussions, I received structured feedback on the various chapters from Chad Woolsey, Grace Shelley, and Victorio Vaz. Colleen Donnelly, a voracious reader, gave book suggestions that usefully furthered my ongoing research. As a next step, several agreed to sit for interviews focusing on their particular area of expertise. I started with Brendan Roark, an expert on oceanography and environmental research, who effectively framed the interactions between environmental research and citizen knowledge. On a very different subject, Joe Gaines, a retired police officer and former military police, outlined the framework for the changes and action needed to create equitable police enforcement. Looking at necessary institutional and organizational change, Summer Roark, teacher and head of a high-school science department, described educational changes in high school systems featuring STEM and STEAM, as they are incorporated into eco-programs to better engage students.

Throughout the book-writing process, I relied on two people to advise me. Thao Ha, sociologist, friend, and fellow traveler on all social justice issues, helped me create the social lens necessary to identify the involved issues. Louis Wilson, renowned African historian and longtime dear friend who has recently left us too soon, was only a phone call away. He was always ready to advise and tell me if I was on the right track, and how the present fits with the past.

More recently, I have been extremely fortunate to rely on the incisive reading and critiques of final chapters by Ann Ferren and Deborah Rhodes, who as friends and neighbors allowed me to rely on their substantive professional skills. Doug Carlson and Alex Nowraiize, IT specialists and also friends and neighbors, were always available to save me from my many almost-computer-disasters.

However, none of the above would have ever come about if it was not for my ever-loving spouse, Phil. During these six years, he has been my advisor, editor, reader, ever-demanding fitness coach, and perhaps most important, chef extraordinaire. For all that, I express immense gratitude and love.

Finally, however, manuscripts don't turn into books which are published without the selection and support of an editor. For that, I would very much like to thank Michael Gibson, Senior Editor at Routledge, and his predecessor, Dean Birkenkamp.

1
INTRODUCTION

The necessity to change *what we do* and *how we live* has now taken on a reality that we did not expect. While many people still hope that our problems will be solved simply with better political policies, sadly, this is not true. Our crises are more profound than simple politics. As a result, we're not sure—and too often divided—on how to move forward. However, one thing is for sure: now and beyond, we all have extraordinarily consequential decisions to make. The first one of critical import is: do we *acquiesce* and shrug our shoulders; *fight* each other to the bitter end; or do we *fearlessly change*? This book is for those who choose fearless change so we can truly learn to flourish together.

We all intuitively understand that too many things have been off-kilter for far too long. Repeated incidents of violence, environmental plunder, profiteering, and injustice resonate from past to present and, if we do not change, into the future. We have tried to make things right in the past and have backed off at critical moments. But we still have another chance. This time, however, we cannot allow our attention to be deflected from the true realities of the situation—and the reality goes far deeper than the simple politics of left and right. Instead, for true change, we need something more comprehensive, more profound. We need a culture shift, a collective change of consciousness.

Changing Perspectives

Of course, culture shifts don't just happen. They require a strong focus on social, cultural, and collective human connection. This is where deep participation and social integrative power come in. Thomas Paine, that instigator and chronicler of our first revolution in 1776, maintained that the real revolution was of the mind: "We see with other eyes; we hear with other ears; and think with other thoughts, than those we formerly used."[1] Once we begin to collectively "see with other eyes," new actions, new practices, and new social institutions suggest themselves. Once we begin to "think with other thoughts," the probability of trust in a just and peaceful nation presents itself. Can political/economic power do that? Clearly not. Culture shifts are not within their realm of power.

But that is exactly what social integrative power and deep participation offer, and it is here that *Fearless Change and Social Action in Difficult Times* situates itself. It illustrates how deep participation and social integrative power—the heart of this book—operate together at that pivotal juncture between the political/economic and social factors of society. If the interface between political/economic power and the largely unrecognized social integrative power gets out of sync, trust is degraded. This emerging clarity makes all the difference. Distinguishing the difference between *political/economic* power, which depends upon trust, and *social integrative power* (also called social power), which creates trust, is essential if new and positive equilibriums are to be achieved. As a result, we begin to understand that political/economic change can only take us as far as existing trust allows.[i]

One way to look at this is by comparing political power with social integrative power in our daily lives. Political power uses "negotiation" which fits perfectly when there is existing trust already engendered by shared values and good governance. But how do you negotiate when institutions and system structures are weighted against you or compromise when the starting points are not equal? As journalist Charles Blow plaintively put it in a TV interview: "How do you meet a bigot in the middle?" The answer is, of course, you can't and you don't: trust is missing. So, while personalized vigilante anger is of no use and actually quite destructive, the calls for negotiation and compromise *are no less useless* when the starting points in reality are starkly unjust. As a result, in these times, negotiation, accommodation, and compromise—the mainstays of political power—*are now rarely effective.*

Certainly, we must hold the perpetrators of injustice and violence accountable, but that is only a part of the process. Malcolm Gladwell, the well-known journalist, describes his antipathy to "blame": "I don't like blaming people's nature or behavior for things. I like blaming systems and structures and environment for things."[2] But blaming people and their behavior is the follow-on to the coercive nature of economic/political power, particularly when it is out of sync with social power. In light of these observations, it is clear that if we

---

i I should note that this discussion will focus on the United States. But because similar problems exist across the world, and initial research indicates that culturally mediated deep participation dynamics work across multiple societies, these same deep participation processes, carefully attuned to society and culture, do work elsewhere.

are to access the cohesive energy of social integrative power, then negative negotiation and blaming are not the answer. When these factors are in the mix, neither collective altruism, empathy, nor inclusion—all essential to the deep participation process—can be accessed. Instead, we would find ourselves, once again, in the toxic realm of distrust, opposition, and exclusion.

There is, however, one critical distinction that each one of us, individually, must make if we are to understand this deeper reality. I was first presented with this distinction more than a decade ago while attending a Buddhist meditation camp. One day in the third summer, I shook my head and said to myself—'this is not for me; I can't do this anymore.' So, I went over to the Lama's office to thank him for all his kindness, but to also tell him that I was leaving. Of course, he asked me to sit down and explain why. I said that oddly enough, I found that everyone was too nice! I further explained that this orientation did not fit me at all. Working as a sociologist where much of my work focused on injustice, I didn't feel that I needed to be nice, sweet, or accepting all the time. So, I didn't fit in and needed to leave.

The Lama looked at me for a while and smiled, and I have to admit, it was an exceedingly kind smile. He finally said: "there are many paths, Paula. There is a path of 'righteous anger' which I believe you might follow. It is a strong one and a possible one, but it is also a very difficult one for any of us. It is difficult because in following it, there is a great possibility that the intended righteous anger can be transformed into *self-righteous* anger by the practitioners themselves. And that is a grave danger."

That idea intrigued me, so I stayed. Over the years since, I have come to see this distinction, and the insights that followed, as one of the most helpful personal and professional pieces of advice that I have ever received. The distinction between righteous and *self*-righteous anger, when observed over time either in ourselves or in the groups around us, creates an awareness of the energy fueling the dynamics at work in both individuals and communities, and what is actually shaping the larger patterns of the culture as well.

When observed, it becomes clear that oppositional blaming dynamics tend to create a self-appropriated and destructive anger—and then, guess what? Here come the toxic, divisive, and polarizing anger vigilantes! This can be avoided, of course, if the individual anger itself is surrounded by a larger cohesive and collective band of community-bound trust and solidarity. But as we all have come to acknowledge, trust becomes quite rare as societies begin to break apart. So, it is certainly right to ask ourselves if a new trustful equilibrium can even be achieved in this era of deep division and political anger.[3]

So, how do we do this? Do we remain 'nice' and 'polite' even in the face of injustice and violence? Absolutely not! The dynamics of *deep participation* and *social integrative power* totally encourage positive 'righteous anger' toward all acts of unfairness, not the perpetrators—but the acts—of injustice, violence, brutality, and hurt. So, in deciding to practice social power and righteous anger, we are definitely not talking about simple politeness or acquiescence here. Instead, we are talking about redirecting our anger about injustice to its root causes. In other words, we can effectively use and even expand solidarity for justice by giving up *self-righteous* anger and using only *righteous anger*. Think about the implications of that statement. It uncovers hidden and critical distinctions and intensifies positive possibilities. When used correctly, righteous anger carries great power for those who utilize its practice.

Most important, when we decide to cultivate only righteous anger as individuals and groups, something different starts to happen—something strong and fire-breathing, but at the same time, heart-tending and loving. We begin a march toward a shared justice that is inclusive from the very beginning—we slowly become the "Includers." Working together within this social approach allows us to call out to the other side and say: "hey, you're going the wrong way; come and join us over here." At first, some of the multiple "Fence-Straddlers" and perhaps a few hardcore "Excluders" will stop and look, possibly attracted by the obvious solidarity of the "Includer" group. Over time, the crossings do grow; multiple Fence Straddlers and, finally, even some of the Excluders begin to join. Exclusion, anger, and violence begin to dissolve as inclusion and social energy begin to expand, and the change that we want really does take place. No, don't stop reading yet—this is not just another "love is all we need" riff.

The Four Ancient Wrongs

Certainly, it is an understatement to say that there are multiple problems facing the world and our country today, but these discrete problems and their potential reforms are not the subject of this book. The complex set of problems to which I refer runs deeper, much deeper. Long-abiding *ancient wrongs* in their current manifestations—existential and interconnected—underwrite our current difficulties. There is a long history of defining and attempting to right these ancient wrongs and it is here each time that the real revolution of the mind begins.

The current manifestations of these ancient wrongs are:

Injustice—Racism, classism, sexism, and anti-LGBTQ remain entrenched;
Poverty—Interlocking structures of poverty produce pockets of dire need in our extraordinarily wealthy country;
Endless War—War and violence undermine even the contemplation of peace;
Environmental Climate Crisis—The existential threat to the planet is morbidly real: with fires, floods, and ocean warming that are precursors of worse to come.

Of course, you, the reader, will immediately say: "But these are the same problems that are being discussed on a day-to-day basis—what's so different?" And I will respond that the difference lies in their definition. Defining them, and separating them simply as 'problems,' as is usually done, dictates a reasoned, political, somewhat superficial approach with various policy solutions. On the other hand, defining them as "ancient wrongs" intimates something more profound. It tells us that they weigh heavily on us in a way that 'problems' do not because, as ancient wrongs, they diminish our humanity. And while they are not intractable, they are well beyond the reach of our usual intellectual problem-oriented methodologies of political and economic change.

All of that is true, and yet the opportunity for deeper, more substantive change does exist. Recent research identifies the link between deep participation, legitimacy, and social change.[4] It illustrates that deep participation and social integrative power, while rarely recognized, specifically work at this more profound and seemingly intractable level. Deep participation, based on the dynamics of connectedness and mutuality, enables participatory groups to review their individual and collectively shared thought-worlds.[ii] Groups can decide which thought-worlds to discard because they are false and which to keep because they are true, and then begin to re-organize themselves within their preferred culture. It is these reinvented organizations and institutions that offer stabilizing support in times of crisis for desired collective culture shifts of thought and belief which then initiate the longer process of social transformation.

This rarely recognized social integrative power, defined as horizontal and connecting rather than vertical and dominating, confers acceptance and legitimacy on new elements, ideas, or institutions within a society. In other words, it gives groups and their societies the actual ability to culturally re-imagine their current guiding images for more positive beliefs that lead to effective social change. Deep participation and social-integrative-power do so through the power of *social relations* which engenders trust—using the strengths of "belonging, social status, identity, love, trust and loyalty" to create a new and shared legitimacy. According to Kenneth Boulding, a well-known economist, "no institution, pattern of behavior, or role structure can exist very long without it."[5]

All of this goes well beyond our understanding of normal modes of change. Social integrative power is not something that we are aware of in everyday life. So, we rarely recognize it when, in unstable times such as these, this

---

ii *Thought-worlds*, a phrase coined by social anthropologist Mary Douglas and others, describes a concept explored by multiple sociologists and anthropologists, beginning with Emile Durkheim. It describes the role of society in organizing social thought, contending the rightness' of some ideas and the "falseness" of others as initially handed out to the individual by society. It does so through guiding images, learned assumptions, and cultural metaphors. In stable societies, increasing individual and shared collective incoherence of inherited negative thought-worlds begins the experience of instability. But changes, in some instances, can initiate a societal *culture shift*.

social power becomes not only useful but necessary. Given our present situation, this power is critically important for four reasons:

- *Social integrative power*, through deep participation, facilitates change and creates cohesiveness and trust at the deepest levels of society and community.
- It is where we *think, believe,* and are *connected* to each other and the world.
- It is where our hearts (*resonating emotion*) and our heads (*critical thinking*) come together and connect.
- It is this trusting *connection* among us which allows groups and societies, particularly in unstable times, to define new shared *social legitimacies* for positive change.

However, if we are to use this deep participatory action process to its fullest extent, we need a broad yet concise definition of the four ancient wrongs' root causes, as well as how to address them. In other words, we may not know or agree on what the *solutions* are, but we must agree on what the *root causes* are if we are all to head in the same direction. That is why the initial focus on the four ancient wrongs is so important. At this moment, we have only a piecemeal idea of what these wrongs are and how they interact and strengthen each other. And without this focus, neither the critical mass for social change nor the necessary maximum impact will be achieved.

'Why' and 'How' Are the Questions

Let me say that this is *not* a book about national and local politics—but *it is* political. It is political because it is about power and how we change things in a real and substantive manner. Recent toxic political leadership at the national level, and the chaotic environment created, has unexpectedly now pushed us to the precipice of real choice. For that reason, let's stop for a moment and reflect. In normal times—in other words, in stable times—the political processes of negotiation and compromise have achieved a great deal of agreement. But this only happens when these processes are working within and are surrounded by the unseen—often subconscious—societal belief systems that are in sync with each other. But now, the out-of-sync factors between the political/economic vs the social/cultural factors are increasingly obvious, and their fraying and breaking apart have been pushed into overdrive. Thus, we begin to foresee the possibilities of a very negative termination point. This is the divisive direction that we as a society and nation have been heading in for some time, but the violent destination remained ambiguous—until recently.

For full transparency, I probably need to add here that recognition of these out-of-sync factors, beginning for me as early as 2015, would not have been so apparent if I had not worked in several conflict-ridden countries under authoritarian rule. So, it is a particularly sad surprise that my experience in conflict-ridden countries is now useful in my own country. It is not something that I ever expected. So, while the 2020 election supposedly gave us breathing space, it certainly did not wipe away the dangers of incipient authoritarianism and the ever-increasing expansion of violence—in fact, by 2024 they had only immensely increased. And we cannot, in these situations, find any foundation for effective negotiation or compromise with those political perspectives which implicitly assumed—and now demand—that some groups of people are to be considered "less than," and are therefore either expendable or not worthy. Nor can we, in these same situations, continue to assume that our immediate individual comfort supersedes other looming problems, particularly the well-being of our rapidly warming planet Earth.

The only choice now is real change or acquiescence. But real change is not just winning the next election. Yes, if Mr. Trump wins that 2024 election, as some believe he might, we would be on a straight-line to autocracy and more violence. But win or not, he is only the tip of that proverbial iceberg. He is not the real problem.

There is a more profound dynamic and pattern in our society that has yet to be acknowledged if we are to achieve real positive change. In every instance, the critical question of "WHY" we as a society—with all of our cherished democratic values—have arrived at this juncture is rarely asked, and never answered. How do we explain, for example, why our current world views no longer fit physical reality (think environmental crisis)? Or why do we stare with incomprehension as it becomes clear that our worldviews no longer fit social reality (think increasing hate and violence)?

Because "WHY" is never asked, it is almost impossible to ask "HOW." How do we fix this? How do we correct this? How do we move beyond our preferred ideologies and even think about this? And equally important, how do we achieve the requisite changes fast? Yes, the reorganization required of us and our society seems to be so profound that the singular actions and processes available to us as individuals are regarded as too insignificant, and perhaps even contributors to our ongoing divisiveness—so we remain frozen and the questions remain.

But no matter their importance, these questions are all dependent on the one question that we do not want to ask or acknowledge. In the following pages, we will explore this question so brazenly lodged in that equality/inequality equation but so consistently ignored, because we don't want to face or acknowledge it. But the questions of "why" we find ourselves in this position and "how" we fix it collectively as a society and culture can only be truly defined through courageously exploring this long-hidden and long-ignored social equation together.

My investigation of the four ancient wrongs, as described here in Part I, clearly indicates that it is the *dynamics* and the *patterns* of inequality that must be examined and acknowledged if we are to understand *why*. In other words, it is all about the *social* underpinnings of culture and society and how the four ancient wrongs inter-relate with mistrust, oppression, and inequality; as well as the inter-twining of each with the other, shaping our society's institutional processes and our individual beliefs in negative ways. When the four ancient wrongs are regarded as a totality, these interactions become quite apparent. Thus, the more profound structures within which we live begin to appear.

I clearly understand, given our increasingly individualized and transactional lifestyle, too often accepting of violence, the skepticism of any power that is described as social. But one reason that deep participation and social integrative power are so appropriate for this era is that their end result can truly reform, reinvent, or reimage our social, political, and economic institutions when just resisting injustice or creating new economic or political policies is not enough. As a result, deep participation instills energy plus solidarity to the formulation of new societal frameworks, adding a necessary measure of positive meaning to our collective action and individual well-being.

Culture Shifts and Thought-Worlds

In today's world, this certainly doesn't mean letting go of the political fight—far from it. Instead, it means recognizing what can be accomplished politically to enhance justice, equality, and democracy. But similarly important, it also means recognizing the areas which require culture shifts. Here, the needed transformative change, as we will begin to understand, can only be accomplished through collective practice of deep participation, culminating in acts of social integrative power. In other words, in these situations, there are no longer any political or economic quick fixes. At the same time, through collaborative participatory work to dismantle negative thought-worlds, culture shifts do take place, sometimes relatively quickly, particularly in these times of great instability. I can assure you, for the culture shift possibility at least, deep participation is the only way!

In exploring all of these new perspectives, *Fearless Change and Social Action* yields critical and unique findings regarding transformative change. It first identifies *how* the current manifestation of the four ancient wrongs always escapes our well-meaning and best efforts to combat them. It explains for the first time, how trust in each other and trust in our system of democracy continues to be degraded by the continuing practice of the four ancient wrongs. To divest ourselves of these ancient wrongs and their current manifestations, *Fearless Change* explains exactly what keeps them in place, why they remain in place, and how to transform them. Through this investigation, we begin to regain stabilizing trust in ourselves and our society.

With this solidarity-growing participatory work, we also begin to explore those out-of-sight, out-of-mind collective *thought-worlds* which give legitimacy to the *implicit* rules of society and, in turn, govern the e*xplicit* rules of politics and the economy. While it is rare that society's implicit rules are recognized in normal times, in unstable times such as these it is different. Society's thought-worlds and their resulting implicit rules, well-anchored in our individual and collective sub-conscious, slowly become apparent. As a result, they are not only more easily recognized but also more easily addressed.

Some thought-worlds are in sync with a just world; some may simply be no longer relevant, while others are downright harmful. For example, still-existing unrecognized thought-world legitimacies—such as inequality and societal structures giving license to racism, classism, and sexism, or similarly, license given to environmental plunderers—have become contradictory and even fatal to our present-day ethical aspirations and living conditions. But still these unrecognized thought-worlds remain. As a result, they provide the sub-conscious scaffolding for the current indifference to truth and the everyday far-too-easy acceptance of lies currently found on social media. Equally important, these negative thought-worlds are contradictory and fatal to the physical and environmental realities necessary for a humane life on a living-world planet. Bringing these thought-worlds forward for examination and understanding can, and must, become an essential—and existential—part of our social repertoire for change.

This is where that unexpected finding that we do not want to acknowledge bursts into full sight. There is a clearly identifiable through-line of *inequality* that runs through each of the four ancient wrongs. We all understand this situation; those ongoing "culture wars" to which we devote so much time and energy are really all about a fight for greater equality—we recognize that. The unexpected finding that we don't want to acknowledge, however, is all about us. Neither the greed, or at least the indifference of the wealthy elite class, nor the brutality of the market-friendly boot-straps class, are the true culprits. Instead, it is also the *social maintenance of inequality* by almost all of us—hidden from sight in the name of comfort and community—which is the main nefarious, stabilizing force. But it is here that negative thought-worlds can be banished and rapid culture shifts can have their beginnings.

Creating Fearless Change

Hundreds, if not thousands, of books have been written about each of the ancient wrongs, their current manifestations, and efforts to combat them. Particularly in the past 10–20 years, the number and quality of books exploring, in depth, one or more of these ancient wrongs has increased impressively. And we are lucky to have them available to us—often for free—at the public library. Each of these books explores in depth a particular ancient wrong. It is interesting to note, however, that many of them end with the same conclusion. Even though each book focuses on a particular ancient wrong's current political/economic manifestation and possible solutions, all of the conclusions feature similar appeals and laments. They all request more dialogue, changes in belief, more solidarity, more social movements, and particularly, new and different worldviews.

This focus is found, for example, in the highly popular books authored by Bill McKibben, Naomi Klein, and Michelle Alexander. McKibben in *Falter,* writing of the environment, identifies nonviolence as "one of the signal inventions of our time … the full sweep of organizing aimed at building mass movements whose goal is to change the zeitgeist and hence, the course of history". Naomi Klein, in her book on the environment, *This Changes Everything,* tells a similar story. "All of this is why any attempt to rise to the climate challenge will be fruitless unless it is understood as part of a much broader battle of worldviews". Michelle Alexander, writing in that enduring classic about the mass incarceration of black people in *The New Jim Crow,* summarizes the situation with blistering candor. "Those who believe that advocacy challenging mass incarceration can be successful without overturning the public consensus that gave rise to it are engaging in fanciful thinking. Isolated victories can be won—even a string of victories—but in the absence of a fundamental shift in public consciousness, the system as a whole will remain intact".[6]

Fearless Change and Social Action, in contrast, sprints fast with minimal detail but answers the requests of the previous books. Its intent is to construct an interactive knowledge base which creates a broad mutual departure point to address those same ancient wrongs. While profound explorations are necessary when exploring one or two subject areas, the "sprinting fast" methodology has a different objective; it clarifies critical connections and intersections among the various subject areas. At the same time, these interconnections traverse a multitude of scholarship perspectives and reveal pertinent evidence-based facts from a number of fields. Once these connections are understood, awareness dawns for the first time. Using social power to review and dismantle the negative thought-worlds becomes no longer just a possibility but a potential reality instead. This is the first step in removing blockages for the development of culture shifts.

This book is divided into three parts: (i) investigation of the four ancient wrongs; (ii) activating social power; and (iii) culture-shift action. Each of the chapters in Part I poses two questions: 'where are we?' and 'how did we get here?' These queries are essential to understand the change of direction that must be taken in order to answer the critical question: how do we change things? While most of us are reasonably conversant with the history and impact of at least one ancient wrong, few of us are adequately informed about all four current manifestations. Therefore, each chapter investigates in some detail the history and impact of a particular ancient wrong and its interconnectivity with the others. This exploration then identifies the enduring wounds to our shared cultural psyche which act as obstacles to change, necessarily moving deeper into that social angst we consistently attempt to avoid.

However difficult for you, me, and all of us, it is necessary to explore the complicities and complacencies, both historical and current, which tell us just *why* and *how* we got to this place. You ask: are you sure about that? Yes. These ancient wrongs are not just terrible individual tragedies or the inevitable sufferings of life. Instead, these ancient wrongs have been integrated into both the *system* and *structure* of American law and culture, putting together the bad with the good. As a result, we must clearly understand the deficits, strengths, and fragilities of the institutions; the structures and the policies of this ancient wrong system in its totality.

We need this information deep-dive even if we only want to usefully reform the system; but more hopefully, particularly if we want to 'reinvent' it. We can't do this, however, unless we understand *why* and *how* this ancient wrong operative system has so strongly contributed to our current world. As a result, the root causes, long well-hidden by negative thought-worlds, are identified—some for the very first time.

The bad news is that while this fabrication by the four ancient wrongs may have been initially constructed by a faceless 'them'; it was, and is, *us* that keeps it running in the multiple cultures of our society. Perhaps it has been unknowing; perhaps it is because we are unconscious as individuals—but still, it is us. The good news is that it is also *us* who can change things. Because these four ancient wrongs are still operating today and because we are living in this society, we have the potential, the gift, and the responsibility to create that "real revolution of the mind" that Thomas Paine so presciently described long ago.

Part II of *Fearless Change and Social Action in Difficult Times* moves us into a series of analytical assessments which are rarely thought about or discussed. Working "below the decibel of politics" at the social power and culture level, we find the thought-worlds which have, for so long, kept the four ancient wrongs in operative force to maintain inequality. This knowledge, in turn, substantially augments a collective "social-self-knowledge of society" that is impossible to attain when working individually at the political and economic power levels. More specifically, this collectively defined social-self-knowledge serves as an analysis base for how the four ancient wrongs have managed to continue operating. It turns out that what we thought was right is often wrong because enduring negative thought-worlds altered perceptions of real-world reality.

The most unexpected finding, as noted earlier, is that neither the greed of the wealthy economic class which liberals tend to believe in nor the corrupting ideas of the current political party that conservatives rely on are the main stabilizers of the four ancient wrongs. Instead, it is the *social maintenance of inequality* by almost all of us, hidden from sight in the name of comfort and community, which is the main nefarious, stabilizing force.

As a result of these findings, Part II clarifies where we are now, how we got here, and why our current worldview no longer fits physical and social reality—whether it is injustice, endless war, poverty, or climate crisis. Based on this assessment, it defines new insights and exit strategies designed to reformulate societal trust and community balance. But the answers are not always what we expect. Learning to more effectively rely on the *collective* and *relational*—including mutual sharing, music, laughter, imagination, camaraderie—is more important than we assume or are accustomed to.

Part III focuses on energizing and far-reaching change realities. It dives into the dynamics and frameworks emphasizing the how-to reinvention issues which are rarely either identified or explored. When deep participation's six elements are practiced, as discussed at length in Chapter 13 it fosters access to social power by numerous self-organized deep participation small groups. Chapter 14 then illustrates how to further organize enduring social movements in the community, region, and nation based on these same small groups.

All of this furthers the legitimacy of new thought-worlds, new directions, and new solutions. Because all levels of deep participation groups are working under the broad banner of the four ancient wrongs, new and self-sustaining long-term social movements begin to emerge for maximum impact. As a result, through choosing multiple venues for reinvention, reimaging, or reform, deep participation with its resulting social integrative power can begin to undo and dismantle the negative systemic structures of the four ancient wrongs. This begins resetting our social order and society for true equality.

Departure Points

I admit to occasional doubts, however. I do sometimes ask myself if the relevance of deep participation and social integrative power will be challenged by the increasing violence that has re-appeared in our society over the past years. As I was finishing rewrites on this manuscript, I read a review about a new book entitled *Grime* that captured the discrepancy between what I am writing about and that more violent version of the world. The reviewer lets the author, Sybylle Berg, tell it as she sees it.

> When you use the instruments of democracy to completely pulverize trust in democracy—that is, put absolutely rubbish individuals in top positions, instigate civil wars, incite the so-called good against the so-called bad by means of Nudging, by the manipulation of their goddamn brains via devices, social media, false information, when you render the press utterly untrustworthy, when you encourage brutality, Nazis, ignorance, and fascism—in short, when you perpetrate insane chaos ....

The book reviewer closes by stating: "no other book has so thoroughly rattled me about where we're headed."[7]

So, yes, I think we are all a bit rattled about the future, with grave doubts on how to move forward. And I completely understand that deep participation and social integrative power may not seem to have the toe-to-toe strength we often assume is needed to counteract this violence. But when I think it through, I realize that social power—the only power that can offer the rebuilding of trust and legitimacy—will outmaneuver bullying forces every time. Violence against violence doesn't win. Deep participation and social integrative power are the *power-tools* that successfully inhibit violence and its chaos as well. So, I have decided to stay on track, writing about this deep-participatory, cultural, historical, circle of social power and helping to create access to these power-tools for all of us.

Rather than discouraging me, this new perspective also helped me realize the importance of distilling a *mutual point of departure* that we could all use in our own creative ways. If we were living in the relatively more relaxed times of societal stability, perhaps each deep participation group could do that on their own. But change is bearing down on us, and we are literally running out of time. Given this reality, we need to maximize our impact as fast as possible. But there is more to this. Many of us feel trapped: our awareness tells us that something has gone terribly wrong, but we don't quite know what it is that we need to do. As a result, we find ourselves, as one writer put it: "stuck within the culture's exhausted concepts and narratives … unable to arrive at a true alternative."[8] This is particularly true for the younger generations under thirty. Many of them intuitively see the reality we as a culture continue to deny—saying one thing and doing another. But while the alternatives offered by politics are absolutely necessary, they are not enough.

So, yes, there is still more. It's all about the things we don't understand, but we know are there. In this situation, the positive force of deep participation and its resulting social integrative power assist us in capturing and deciphering those amorphous meanings and concepts floating just beyond our grasp. Building awareness of negative thought-worlds and consciously putting them aside allows us to move forward—sometimes in the dark at first—to discern new positive ideas and make them into realities. It does take courage, but this is the way better futures are created.

Notes

1 Eric Foner, *The Story of American Freedom* (New York: W.W. Norton, 1998), 16.
2 Dan Amira, "Malcolm Gladwell Likes Things Better in Canada" (*Talk* column interview, New York Times Magazine, June 3, 2018), 62.
3 Steve Rattner, "Charts on Growing Divisions Among Americans", Sept. 6, 2023, https://www.msnbc>morning-joe

4 Paula Donnelly Roark, *Social Justice and Deep Participation: Theory and Practice for the 21st Century* (England: Palgrave MacMillan, 2015), 15. This book defines deep participation for the first time, based on two prior research initiatives, discussed in Chapters 7 and 12.
 5 Kenneth E. Boulding, *The Economy of Love and Fear* (California: Wadsworth, 1973), 5. Further discussion of social integrative power is based on *Three Faces of Power* (London: Sage, 1990). Also based on extensive discussion notes from PhD economic classes given by Professor Boulding.
 6 Bill McKibben, *Falter: Has the Human Game Begun to Play Itself Out?* (New York: Henry Holt & Co., 2019), 219.
 Naomi Klein, *This Changes Everything* (Simon & Schuster, 2014), 460.
 Michelle Alexander, *The New Jim Crow: Mass Incarceration in the Age of Colorblindness* (New York, : New Press, 2012), 234.
 7 Ron Charles, "A caustic satire burning up the future", review of *Grime* by Sybylle Berg, *Washington Post Book Section*, Dec. 18, 2022
 8 Daniel Oppenheimer, "Out of unbearable loss, a vision of radical hope", *Washington Post*, Book World Section, Review of *Imagining the End* by Jonathon Lear, Nov. 13, 2022.

Bibliography

Alexander, Michelle. *The New Jim Crow: Mass Incarceration in the Age of Colorblindness*. New York: New Press, 2012.

Amira, Dan. "Malcolm Gladwell Likes Things Better in Canada", *Talk* column interview, New York Times Magazine, June 3, 2018.

Boulding, Kenneth E. *The Economy of Love and Fear*. California: Wadsworth, 1973.

Boulding, Kenneth E. *Three Faces of Power*. London: Sage, 1990.

Donnelly Roark, Paula. *Social Justice and Deep Participation: Theory and Practice for the 21st Century*. England: Palgrave MacMillan, 2015.

Douglas, Mary. *How Institutions Think*. New York: Syracuse University Press, 1986.

Durkheim, Emile, Carol Cosman, trans. *The Elementary Forms of Religious Life*. New York: Oxford Classics, 2008.

Foner, Eric. *The Story of American Freedom*. New York: W.W. Norton, 1998.

Klein, Naomi. *This Changes Everything*. New York: Simon & Schuster, 2014.

McKibben, Bill. *Falter: Has the Human Game Begun to Play Itself Out?* New York: Henry Holt & Co., 2019.

Oppenheimer, Daniel. "Out of unbearable loss, a vision of radical hope", *Washington Post Book World*, Nov. 13, 2022.

Paine, Thomas, *Common Sense*. New York: Penguin Books—Great Ideas, first published, 1776, 2004.

Rattner, Steve. "Charts on Growing Divisions among Americans", Sept. 6, 2023, https://www.msnbc.com

PART I
Investigating the Four Ancient Wrongs

2
INJUSTICE: THE MALICE OF RACISM

All of us recognize the first few lines of the Declaration of Independence, and many of us know it by heart:

> We hold these truths to be self-evident- that all men (sic) are created equal, that they are endowed by their Creator with certain unalienable rights, that among these are Life, Liberty, and the pursuit of happiness.
> That to secure these rights, Governments are instituted among men, (sic) deriving the just powers from the consent of the governed.

So yes, we must thank our founding fathers for getting the ball rolling with that beautifully written Declaration of Independence and their courage in establishing a new nation under new principles. But who must we thank for insisting that we live up to those explicit words of freedom and equality in the Declaration? That's right: the Native Americans who were consciously slated for extinction; the enslaved Africans who were regarded as chattel property; the so-called benighted Irish and all other immigrants of the previous centuries; the women who volunteered and staffed the underground railroad; the female suffragettes who insisted on both equality as well as the vote; and now the brown and black folks—the dreamers—who come across our borders seeking security and freedom.

American history has from the beginning twisted and turned, addressing racism in a flimsy, if non-existent, manner—except, of course, in our minds. The abiding tolerance of historic and current racist practice; the erasure of the most egregious and violent acts of racism in our history; our acceptance of "race" as a bona fide category of nature as opposed to a made-up fetish;

and the outrageous trampling of our Declaration's equality promise is there for all of us to see—if only we will.

Many of us will immediately shake our heads in disagreement and say: "absolutely not, I have great sympathy for everyone that suffers from racism and inequality." But as we will see, neither sympathy nor empathy is enough. To get ourselves out of the problem category requires not only 'acknowledgement' but also 'commitment.' Martin Luther King's definition and categorization of the problem was right more than 50 years ago and remains so today.

> Loose and easy language about equality, resonant resolutions about brotherhood fall pleasantly on the ear, but for the Negro there is a credibility gap he cannot overlook. He remembers that with each modest advance the white population raises the argument that the Negro has come far enough. Each step forward accents an ever-present tendency to backlash.
>
> This characterization is necessarily general. It would be grossly unfair to omit recognition of a minority of whites who genuinely want authentic equality. Their commitment is real, sincere, and is expressed in a thousand deeds. But they are balanced at the other end of the pole by those unregenerated segregationists who have declared that democracy is not worth having if it involves equality …. The great majority of Americans are suspended between these opposing attitudes. They're uneasy with injustice but unwilling yet to pay a significant price to eradicate it.[1]

Of course, many of us who are white want to quickly place ourselves into Dr. King's "minority of whites who genuinely want authentic equality." But we must remember, it takes "deeds" not just sympathy. And we must always remember, using Dr. King's words: there are those "who have declared that democracy is not worth having if it involves equality." And lest we forget, the great majority of us, as "Fence-straddlers," have not yet come to terms with the role white people have played in terms of racism.

Given these circumstances, foundational to our society and culture, I believe the best place to start a discussion about racism in America is not about black, brown, Asian, or Native American peoples but rather with white people.[i] Because when you get right down to it, we white people, or, as the American author Ta-Nehisi Coates reminds us, those of us who *"think we are white,"* have been the real problem. So, let's look at where we are and what eradicating racism actually means when called for in multiple protest movements—but still not enacted.

---

i I do not capitalize the words black, brown, or white as I do not wish to legitimize the made-up category of "race," a domain which is now scientifically disputed.

Those of Us Who Think We Are White

It makes sense to start our inquiry by interrogating the unspoken but very real status of 'whiteness' in the United States today—rather than black, brown, red, or yellow—and how that status has changed or not changed. To do so, however, means that we first must recognize the almost unrecognizable: our unconscious acceptance of the ubiquitous and unacknowledged 'whiteness' of our culture. Years ago, Toni Morrison, the celebrated author, offered a particularly insightful example of this unconscious perspective. She pointed out that when talking about citizens of South Africa, it is normal to indicate categories—white South African, black South African, "colored" South African—otherwise we would not know their background.

But, as Morrison tells us, when we speak of "Americans," there is no need to lend such an indicator to our conversation; instead, it is always assumed that we are speaking of a white person *unless* we are speaking of someone who is different from the accepted norm of whiteness, and therefore they must be identified as such. Morrison describes this "as a metaphor for transacting the whole process of Americanization, while burying its particular racial ingredients" and suggests that this essence may be something that the country cannot do without.[2]

So, instead of looking at how far people of color and those of differing origins have come in their efforts to achieve freedom and equality in the last several hundred years, let us start by looking at white people and how far we have to go before we can say that the United States is no longer—either overtly or subliminally—a racist society.

For myself, I have to admit, I was raised to believe, as many of us were, that racism was an aberration in our modern society which was fast disappearing. Slavery (1619–1863) was a thing of the past; Jim Crow (1863–1945) a time of "segregation, disenfranchisement, and lynchings" (so described by Martin Luther King) was finally and decisively over, wasn't it? We did not treat Native Americans well, but they now have multiple reservations with which they are happy, correct? And it seemed certain, with the beginning of the 21st century, our country would see the end of racism. Right?

No. Our current status—as illustrated in simple Census data—proves even those beliefs are not only decidedly wrong, but deadly so, and our present days of confusion continue. In fact, our situation has become so difficult, and we are so unprepared, that it makes one remember the motto—"to save the soul of America"—that Dr. King's Southern Christian Leadership Conference took for itself when it began in 1957.[3] Did any of us, as so-called white people, understand at that point how appropriate that motto was? Probably not; but now, if we look fearlessly at the situation without trying to protect ourselves, awareness begins to dawn.

To better understand the racism situation, our first impulse is to review basic economic and social quantitative data concerning the individual status of white, black, Hispanic, and Native American groups in our nation. Household income, housing, education, and poverty rates give a brief picture of what most people assume is relative well-being. The median adjusted household income in 2018 was $68,145 for whites, $40,258 for blacks, and $50,486 for Hispanics, while for Native Americans in 2017, it was $39,719. In 2018, the official poverty rate for the United States nationally was 11.8%.[ii] Compartmentalized, the poverty rates are: for whites—8.1%; for blacks—20.8%; for Native Americans—39% on Reservation and 26% off Reservation. Asian groups have the highest incomes, with the top 10% at $133,529. This compares to whites within the same top 10% at $117,986, blacks at $80,502, and Hispanics at $76,847. But also, Asian ethnic groups in the United States have the highest level of income inequality due to disparities within various Asian ethnicities.[4] All of this is pre-Covid. Between May and September 2020, at least 8 million more people fell into poverty. However, poverty actually decreased for children during the Covid pandemic due to increased assistance, but is now once again rising.

One of the steadiest measures of our current status is household net worth. According to US 2022 Census data, white households were worth a total of $1,322,528; black households were worth a total of $340,559; Hispanic households were worth a total of $323,682; and "other" households (grouping several together) were worth $872.694.[5] Of note, new Pew Research Center data indicate that over the past five decades, the American middle class has shrunk considerably, from 61% in 1970 to 50% in 2021.[6] It's complicated. This 'shrinkage' is marked by an increase in the upper-income tier as well as an increase in the lower-income tier. For a country that depends on a large middle class for stability, and even its legitimacy, these are troubling markers.

These numbers, although revealing in their disparities, are national in nature. They do not communicate the substantive differences found throughout the United States—from poor Native American reservations to differences within any one particular city. For example, given the tragedy of George Floyd's brutal murder in Minneapolis in 2020, what is the statistical backdrop to this tragedy? How does Minneapolis compare with the rest of the country in terms of overall black citizen well-being? According to NPR's "Planet Money," this placid, comfortably middle-class Midwestern city has "some of the most abysmal numbers on racial inequality in the nation." Black poverty, at 25.4%, is *over four times higher* than the white poverty

---

ii I use the 2018 US Census data as the best representative of enduring patterns. Pre and post-COVID years are not as representational.

rate of 5.9%—both outside national averages. The median black family earns $38,178 annually, while their white counterparts have a family income of $84,459 a year.

Only one city in the country—only one—has an income equality gap that is worse, and that is nearby Milwaukee, Wisconsin. Other differences in Minneapolis add to the dismal picture: in 2019, black incarceration in the city was *11 times* that of whites; in 2016 the black unemployment rate was *more than three times* the white employment rate. NPR further reports that "racial covenants made it hard for blacks to become homebuyers and live in white neighborhoods," and the state itself ranked 50th in terms of racial disparities in high school graduation rates. This then, is the underbelly of the desperation that we all witnessed beginning in May 2020 following George Floyd's murder and other related incidents against minority citizens.[7] A recent report in June 2023 concerning the status of policing in Minneapolis by the Department of Justice underscores the grave injustices suffered by black people.[8]

But what have white people traditionally thought and believed about racism, status, and poverty in the United States? Maybe with the experience of the pandemic perceptions and the expanding understanding of police brutality, it might begin to change, but maybe not. A 2016 Pew Survey of white and black opinions gives some basic insight on these longstanding perceptions, and a more recent 2019 Pew Survey reveals little change. On the question of whether the United States "has made the changes needed to give black equal rights with whites," 38% of whites believed those rights to be already accomplished, while another 40% believed "our country *will eventually* make the changes." In contrast, only 8% of blacks believed the United States has already made the changes necessary; 40% hopefully believe that the changes will eventually be made, while a large percentage—43% of blacks—believed that the necessary changes *will not* be made.[9] Little had changed by 2021 in these numbers. There was only a 1% increase in the number of blacks who believed that necessary changes will never be made. However, 65% of blacks did state that the increased focus on racism had *not* helped their cause.

Assessments of the nature of racism also differ dramatically between whites and blacks. On the issue of racial discrimination, the majority of whites point to individual prejudice, while blacks discount this factor and point to societal and institutional racism (e.g., the education system, the justice system, the health and housing systems, as well as the underlying economic systems and political structures). Only 36% of whites believed these points of racial discrimination to be an issue in 2017 or before, while 70% of blacks hold this totality of racial discrimination to be a major factor. In terms of financial disparities, a plurality of white people (47%) said that blacks are worse off financially; a robust 37% believed blacks are about as well off as whites; and 5% maintained blacks are doing better than whites."[10] In *The Color of Law*, Ricard Rothstein discusses perhaps the most damning statistic of all; as reported in an Atlantic magazine article, it indicates that "*64 percent of white Americans* think the legacy of segregation is either a 'minor factor' or 'no factor at all' in today's white-black wealth gap."[11]

Are you saying to yourself—OK, I get it, can we stop now? The answer is No, not yet. Stopping at this more superficial level is not what we need to do. We now know a bit about where we are as a country, but nothing about how we got here. What all of this means so far is simply that none of us so-called white people can excuse ourselves from this equation. Clearly, the supremacy of 'whiteness' remains an unstated but very real assumption for a major proportion of the US population. If we were not convinced of this reality before, 2020 should have convinced us—54% of all whites voted for Trump. So, our claims that the individual diverse friendships we may have, or even the absence of any personal animus, are sufficient to relieve us of the responsibility to see racism for what it is, are not sufficient. To have true change, the underlying connections and interactions of institutions guiding society, and to some extent our own interactions, will need to be recognized and acted upon. Then can we stop.

Racism Up Close

Now that we have taken that objective assessment of where we are as a society, let's look more closely to better understand the full force of what these statistics indicate. Sadly, some of these unacknowledged interactions are there simply for the looking, and they are incredibly dangerous—to the black, brown, red, and yellow people involved, and to our country as a democracy. One of the most obvious dangers to men, women, young boys, and democracy itself is the violent, indecipherable, and unaccountable killing of black men and women by police officers.

Because these police officers are sometimes not even charged or found *not guilty* in the most duplicitous circumstances, it is clear they have not been held to the bar of equal justice demanded in a democracy. Yes, finally, in the spring of 2020, after the police murder of George Floyd, protest movements hopefully began real change, and in April 2021, the policeman who killed George Floyd was found guilty of murder, bringing both relief and grief to a traumatized family and country. But this murdering and killing obscenity has been going on for a long, long, time, and the question is, will these ongoing protest movements actually go the necessary distance for real change? The Washington Post, for example, reported in early 2023 that these police killings have continued at the same rate despite multiple protests begun after the George Floyd murder.

Have we ever had a real, honest, and sustained discussion, of what that "equal bar of justice" might be? The answer is No, but positive actions are beginning in some states. Certainly, the "rules of engagement" which police are bound to follow concerning killing potential assailants are, in most states, incredibly lax. In California, for example, the enforcement law has been on the books since 1872. It states that a police officer may lawfully kill someone "while arresting persons charged with felony, and who are fleeing from justice or resisting such arrest."[12] Does that leave an open door for the police to do anything, at any time, for the most minor of excuses? I would say so. The ACLU reported that in 2017 California police killed 172 people, more than two-thirds of whom were people of color. A number of these shootings were highly disputed and featured in national news stories, primarily because the young men killed were unarmed. Over the years in California, there have been addendums to the enforcement law, but only in 2019 did a movement begin to re-write the law itself.

A new California Legislature bill, in effect as of January 1, 2020, now establishes clear and limiting standards. The new law allows law enforcement officers to use deadly force only when "necessary"; "when their life or the lives of others are in imminent danger and when there is no other alternative to de-escalate the situation, such as using non-lethal methods." There was initial resistance to the toughening of these rules, many citing the split-second decisions often required of police. But prior positive examples of this kind of policing, as exemplified in Seattle, Washington, convinced the majority of both legislators and lobbying groups to move forward.

The ACLU says the legislation is historic: they noted that it makes "California the only state to combine the 'necessary' standard, along with a requirement that the Courts consider an officer's conduct leading up to the use of deadly force when determining whether the officer's actions were justified." Law enforcement groups also praised the new law. Three of the groups spoke of the legislation—when coupled with a second bill on use of force training—as "landmark legislation."[13] But, as we are all too well aware, the killings across the country continue and have recently become ever more brazen.

Tyre Nichol's murder in February 2023, for example, was similar to Floyd's and others, in its sick viciousness, but it differed in two distinct ways. The first was that the police assailants themselves were black. But as one Memphis citizen characterized the killing: "it was not the color of the officer; it was the color of who is being policed" that mattered.[14] In other words, it is brutal police culture that is called into question. But digging deeper, there may be more to it. Equally, if not more interesting, the second reason that Tyre Nichol's murder differs is that, from all reports, it may not have been intended as a killing, only a violent beating. Several black men, days after Tyre Nichol's murder, reported they too had been violently beaten by the same squad of police.

This potentially suggests that a brutal beating was the intent of Mr. Nichol's death. This possibility brings to light an under-reported historical aspect of police brutality. Recent reporting by Sarah Schulman underscores these factors in her review of three books published in 2021 and 2022, which dig deep into a little-discussed topic: the *history of policing*.[15] These books take us into a historical backdrop offering "historical genealogies for the violations of contemporary policing," which is essential to the more profound understanding needed by all of us. For example, in *The Streets Belong to Us*, A.G Fischer, and A. Lvovsky in *Vice Patrol*, both argue that "vice," focusing on sexual behavior, including public movements of women and queers, was *foundational* to law enforcement.[16]

They explain that more than 100 years ago, when cultural rationales changed from containing prostitution to abolishing it, more than 200 red-light districts were shut down by city governments. But the police, accustomed to payouts from the brothels, gambling, and drinking associated with the red-light districts, did not abolish them as instructed. Instead, they moved these previously established red-light districts into black neighborhoods. Fischer reports: "Once vice had been relegated to Black neighborhoods, those neighborhoods became ever more associated in the white public imagination with 'sexual deviance and lawlessness,' which helped justify the criminalization of their residents."

Continuing this historical backdrop, the authors observe that the 1930s Prohibition failure left police forces completely discredited. As a result, the police "rebranded themselves as 'crime fighters' charged with confronting not wayward girls, but manly gangs, even though this didn't reflect reality." But the police continued their sexual policing—prostitutes were arrested but no johns; and black neighborhoods continue to be tied to criminal behavior.[17]

The Tyre Nichols murder was committed by a SWAT-type team organized to operate in these earlier criminalized areas, now designated as high crime. Given that situation, the history and facts of Mr. Nichol's murder indicate something further must be considered. It should be at least contemplated that these present-day police teams, white and black, operate so brutally in these neighborhoods because of the aforesaid historical initiatives by the police which established the lie of criminality, along with the silent, sometimes unconscious, permission of white society. As a result, have these police, no matter their color, also been infected with the imagined association that *any* black person in any such designated area had to be, at minimum, up to no good and therefore deserving of …?

This real-time history is just one more lesson in how we got here, providing a more nuanced backdrop for better understanding the violence that currently pervades policing and how it affects our society. So, it must be considered: does the criminalization of black people just because they are black and therefore deserving of vicious beatings as a deterrent to their *next* crime become not some unreasonable idea; but rather an astute possibility? Mariame Kaba, author of the third book on policing, *We Do This 'Til We Free Us,* offers alternatives for thinking differently. She has spent her career asking what would happen if we, as a public, understood and opposed abusive policing and demanded that we stop arresting and jailing people as the only deterrent, and instead offered them the necessities needed for a fully human life.[18] What would life be like for all of us then?

But we tend to dismiss these possibilities. Certainly, before instant citizen photos and videos came on the scene, many white people tended to assume the criminal intent of the *victim,* automatically exonerating the perpetrator. With video images, that is no longer possible—there is now recognition that racism and fear, particularly of guns, as well as anger, can be major factors for brutalizing police initiatives. But there are also other unacknowledged issues that hover in the background.

What about the racist constructs of "whiteness" which have helped to cover up the long and shameful history of murdering violence that we white people would like to forget? It is now well documented in a not-so-distant era that at least a thousand black men were lynched every decade—beginning in the 1880s—despite the fact that America's first systemic series of social movements against inequality was in full swing. These lynchings continued into the 1930s, and at least 5000 people were lynched during that period. Carol Anderson has done the difficult research for us.

> By 1920, in fact, there had been more than a thousand lynchings per decade; and in the rebel South, almost 90 percent of those killed were African American. Five states—Mississippi, Georgia, Texas, Alabama, and Louisiana—accounted for more than half of all lynchings in the nation. One of the most macabre formats for the murders was a spectacle lynching, which advertised the killing of a black person and provided special promotional trains to bring the audience, including women and children, to the slaughter.[19]

The majority of Americans have believed that this entire violently macabre era was over and done with—finished! But even before Breonna Taylor's murder at the hands of the Louisville police while sleeping in her bedroom, or George Floyd's public execution, unaccountable police murder of black people had been there for all of us to see—if only we would. And we can't let ourselves forget that it took a reminder by the then rapidly organized "Say Her Name" organization for the public at large to even focus on Breonna Taylor's murder—a most ghastly reminder that the murders of women of color or poor white women are often not counted or reported upon.

And what about the Southern Poverty Law Center's (SPLC) annual hate group count and analysis? It reported a record high of hate groups in 2018 and the fourth straight year of hate group growth. SPLC reports there were 1020 of these hate groups operating across America by 2019—"a 30% increase roughly coinciding with Trump's campaign and presidency." This increase followed a three consecutive years of decline near the end of the Obama Administration.[20] The Anti-Defamation League also specifies that the "overwhelming majority of fatal attacks by extremists in the United States were perpetrated by right-wing domestic extremists." As a result, the Director of the SPLC's analysis group suggested that the continued growth of hate and violence indicates that then President Trump was not "simply a polarizing figure but a radicalizing one."[21] In 2021, with Trump out of office, the SPLC then mapped a decline, with 733 identified hate groups in the United States.[22]

So, a question for all of us: we thought that the racist perceptions motivating those past heinous lynchings had been wiped away. It simply did not seem possible that such massive lethal perceptions could be still lurking in the background and still somehow involved. However, must we now admit that the racist cast with which we have flooded our history somehow continues to interact, often unknowingly, with our actions, even today? Is it possible that this malignant, malevolent, and retroactive racism has begun to metastasize once again in our society? That is a possibility.

Crimes of Being

But all of this remains simply facts and questions. None of this can speak to the entirety of pain and destruction *for* all of us—*of* all of us. No, it's not near enough, so let me borrow someone else's words once again. In *Crimes of Being* Thomas Chatterton Williams interviews John Edgar Wideman, the author of *Writing to Save a Life*. Wideman's book starts with the brutal murder of 14-year-old Emmett Till in the Mississippi summer of 1955; it was this murder that is credited with giving nationwide impetus to the civil rights movement. But for Wideman, it moved him to explore more deeply the unknown and difficult case of Emmett Till's father and, more generally, the isolation of black men in America.

Williams observes that the most unsettling argument that Wideman makes in this book about Emmet Till's father, Louis, is that whether guilty of any particular actions or not, "Till's one true offense is something that can only be accurately described as a crime of *being*: In the logic of the criminal justice system, people like Till, people bound to the wrong side of that stubborn fiction of race, often seem to necessitate a 'pre-emptive strike.'" Chatterton Williams goes on to observe in a 2017 article:

> I had never thought of nor seen Louis Till before Wideman painted him so exquisitely, and now I have to acknowledge he is all around me. Walter Scott? He's Louis Till; so is Eric Garner. Michael Brown, unsympathetic as he appears on that convenience store video—I can no longer see him

without conjuring Emmet's father. Seventeen -year-old Laquan McDonald, wandering through the Chicago night until his body jerks and jumps from 16 shots? Louis (Saint Till). Poor Philando Castile—pulled over at least 49 times in 13 years before the final and fatal interaction that left him bleeding in front of his girlfriend and her daughter and all the rest of us on Facebook Live—is a high tech Louis Till. Ditto Alton Sterling down in Baton Rouge, Freddie Gray up in Baltimore and "bad dude" Terence Crutcher out in Tulsa: all these men are Louis Tills. Trayvon Martin and 12-year-old Tamir Rice are something else altogether, heart-rending combinations of both Tills, "pere" and "fils," doomed man-children in the fretful, trigger-happy imagination of American vigilantes and law enforcement ... these were all crimes of *being* before they were anything else.[23]

So, I must ask, and I believe you must also ask: are these "crimes of being"—if perpetrated by those believing themselves to be white—simply allowed by the rest of us as a society, because the other—the victim—was supposedly *not white*? Or at least people who were standing with people who were *not white*? Perhaps we just did not pay sufficient attention? Could that be it? We tell ourselves that we remain individually blameless—"only those bad apple police that shoot first, ask questions later, or those ICE folks that take things too far (US Immigration and Customs Enforcement), or those that look the other way when injustice is being carried out in clear sight—only these people carry the blame!" we say, we insist ... However, if we truly consider the situation, we know that's not really true, is it? But how is it not true? And how do we change it?

Justice Ignored

The structural and institutional foundations of racial injustice in this country are easy to ignore, at least for those who consider themselves to be white. However, this same foundation is one that black, red, brown, and yellow people come to know with their gut—with their entire being. But all of us, of whatever color and background, tend to wall off the feelings of horror if we can. Some of us do so to maintain indifference, while others of us seek only some distance so we can continue to function. Equally important, even for all of us white, black, brown, yellow, and red people who clearly ascribe to the idea that structural racism is real, true awareness is still not there. Otherwise, we could not continue to ignore, accommodate, and live with the reality of the situation. But even for those who truly see these structures, our current politics remain an insufficient response, and the way forward remains unclear.

But there are books, as noted in the Introduction, that begin to outline the hidden structures of racism and injustice of the ancient wrongs; they and others point to intersecting structures that support the ancient wrongs as an inter-related whole. As a result, the *through-lines* of the four ancient wrongs slowly begin to come into focus. If you, the reader, wish to have greater depth, hopefully you will rush to read—or reread—these books. But in the meantime, starting with the first ancient wrong, I use the excellent research found in these books to outline and better understand how we got here and to more clearly designate the underlying structures of injustice.

In 2018, a number of books commemorated the fiftieth anniversary of Martin Luther King's death. Annette Gordon-Reed, professor of history at Harvard—using both her personal experience and her professional vantage point—chose five of these books for review.[24] Despite the numerous known and unknown aspects of King's life described by the various authors, it was, for me, two of Gordon-Reed's own observations that deserve our extended consideration. Her first observation was how "comfortably" King now fits into our history; Gordon-Reed tells us that MLK and others like Harriet Tubman and Rosa Parks "have now become 'safe.'" This perspective evolved even though there was considerable "white contempt" for King during his lifetime. She observes: "King's elevation to something like sainthood has obscured the truly herculean effort he put into what was called 'the struggle.' *The true nature of his labor has been lost.*"[25]

I am a long-time reader of MLK because of my interest in justice and social change, but I had never thought of his work, or all of the others mentioned, in that particular way—essentially, "labor lost," therefore "stripped of their radicalism." As a result, I immediately went back to one of my favorite books, *A Testament of Hope: the Essential Writings and Speeches of Martin Luther King, Jr.,* edited by James W. Washington. Reading it again from this new perspective, I found that there were numerous examples of the "true nature of labor lost."

Martin Luther King is so indelibly *right* for this moment—he defines the chasm—the cleavage—between blacks and those white people not yet ready to accept true justice and equality. MLK first describes how the Civil Rights movement foundered so quickly after the high point of the 1965 Voting Rights Act. The simple (and white) explanation—"white backlash was created by the Watts riots and Black Power slogans." But King describes and documents how the change in support came about well before these events. Instead, the situation, he tells us, was more complex but also more hopeful in the long run. The Voting Rights Act definitively signaled the end of the first phase of the civil rights revolutions for black Americans. But for the majority of white Americans, it was quite different:

> The first phase had been a struggle to treat the Negro with a degree of decency, not of equality. White America was ready to demand that the Negro should be spared the lash of brutality and coarse degradation, but it had never been truly committed to helping him out of poverty,

exploitation, or all forms of discrimination. The outraged white citizen had been sincere when he snatched the whips from the southern sheriffs and forbade them more cruelties. But when this was to a degree accomplished, the emotions that had momentarily inflamed him melted away .... When Negroes looked for the second phase, *the realization of equality*, they found that many of their white allies had quietly disappeared.[26]

King then asked the question: "why is equality so assiduously avoided?" He gave us his answer more than 50 years ago—a momentous but currently forgotten insight. The majority of white Americans, while ready to adopt basic decency—foregoing brutality and degradation—were not prepared to enter into the second phase of civil rights—*equality to end poverty for all*. "Thus America, with segregationist obstruction and majority indifference, silently nibbled away at a promise of true equality that had come before its time."[27]

We don't remember Martin Luther King for that challenge, do we? So far, we have preferred the sweeter, more comforting quotes. So is this why those who voted for Trump, the 45th president, claim they are not racists? In other words, they truly agree that "the Negro should be spared the lash". But further than that …? So is it time for all of us to show up as true allies to that second challenge of civil rights? Clearly yes. But to do it well? We need to understand more.

* * *

Stamped From the Beginning: The Definitive History of Racist Ideas in America is a 511-page compendium of information that clarifies not only our beginnings as a country but also allows us to better understand our current status.[28] There are three threads of thought that the author, Ibram X. Kendi, makes central to this book, and which begin to unpack our own themes. First, he amplifies the message—discussed for more than 30 years—that our society is *not* necessarily moving away from racism toward justice. Currently, we may be at an inflection point—deciding to definitively move toward anti-racism, or perhaps not. But Kendi makes it amply clear that the evolvement of both racist and anti-racist ideas has marched hand-in-hand together throughout our history.

Narrating the entire history of racist ideas, Kendi starts with their early beginnings in the 15th century. He organizes a large and engrossing amount of material for the reader's consideration and uses well-known historical figures to represent the five eras of racial thought in America. They include *Cotton Mather* (1663–1728) who represents the first era as a staunch segregationist who convinced churchgoers that slavery was acceptable to God.

Thomas Jefferson (1743–1826) represents the second era as an assimilationist: drafting the egalitarian Declaration of Independence but also continuing to utilize slavery for his own benefit. *William Lloyd Garrison* (1805–1879) led and truly believed—with great fervor—in the 19th-century movement for the abolition of slavery. But Garrison continued to use assimilationist ideas, promoting the idea that black people were still inferior but would, in time, become the equal of whites. In the meantime, they were human as god's children, and deserving of assistance and support.

Moving into the 20th century, W.E.B. Dubois (1868–1963), representing the fourth era, was known as "the first professionally trained Black scholar." He initially adopted popular assimilationist ideas, as did many other black people at the time. But Dubois came to challenge, with the rise of "Jim Crow," these ideas and developed era-changing transformative and influential anti-racist ideas. Angela Davis (1943–present), the eminent California professor, represents this current era as its tough and strategic leader, fighting to maintain the racial progress of the 1960s against new producers of white supremacist thought.

Kendi's narration makes a critical distinction. It's one that I am grateful for, as I did not clearly understand all of the necessary nuances. Perhaps others of you may find it similarly useful. He points out that from even before our Declaration of Independence, up to our debates of today, there have always been *three* sides to this racial conflict—not the two that most have focused upon. Certainly, the racist *segregationists* are known to believe that black people are inferior to white people, while the *anti-racists* believe black and white people to be equal. We also knew, as Kendi points out, the role of the *assimilationists* who believed that the only way for black people to become equal is to first become worthy and assimilate into white culture.

But Kendi goes further. He explains how it began, and often still continues. Concerning our historical leaders: "We have remembered the assimilationists' glorious struggle against racial discrimination, and tucked away their inglorious partial blaming of inferior Black behavior for racial disparities."[29] The most prominent assimilationist of them all, Thomas Jefferson, is, of course, famous for the stirring words found in the Declaration of Independence. At the same time, all of us also know that Jefferson kept his workers and servants, including his mistress, Sally Hemings, and the six children he fathered with her, all enslaved.[30] On the other hand, perhaps we judge Jefferson too harshly when we consider that related actions continue today. There are far too many—from politicians to historians to business people—who continue to practice the same type of self-interested thought and action and don't even recognize it. As Kendi points out: "Racist ideas have done their jobs on us."[31]

But, according to Kendi, there is more to it than that—and this is one thing I did not sufficiently understand. Assimilationists also conferred a societal *dispensation* upon all of us for the future by asserting that the belief in the inferiority/superiority of difference was simply "human nature"—that's just the way it is. When I thought about it, this perception explains a lot. The *assimilator* belief has clearly diminished, but the *Fence-straddler* role has been galvanized and expanded—that "dispensation" gives difference and inferiority an odd staying power. That straddler can either jump the fence to fight for a right that benefits him; or he can lounge back, shrug and say—"what can you do, that's just the way it is."

In other words, 'Fence-straddlers' still believe in 'difference,' but they practice it in a relatively new and effective form of neutrality or apathy—best illustrated through their "indifference" and lack of care for others—for which they are not condemned. Although lighter in tone and intent, this practice still gives our country and all of us individually a built-in excuse for failing to come to terms with both racism and the totality of the four ancient wrongs. It is this current variation—from segregationist to assimilationist and now to the popular fence-straddler role—which is particularly successful in muddying true understanding of racial injustice.

* * *

Benjamin Madley's *An American Genocide: The United States and the California Indian Catastrophe* (a 2016 Los Angeles Times Book Prize winner) delineates the hidden structures of racism *and* whiteness in a different situation. Although the country's inherent belief system in our "exceptionality" excludes the possibility of US genocide, Madley proves that the facts indicate otherwise. According to Madley, this annihilation of the California Native Americans needs to be remembered for what it is.

> [T]he near obliteration of its indigenous peoples remains one of the formative events in the nation's history. As in many other Western Hemisphere countries, the Native American population cataclysm in the United States played a foundational role in facilitating the conquest and colonization of millions of square miles, the real estate and natural resources on which the country was built. Thus, how we explain the Native American catastrophe informs how we understand the making of the United States and its colonial origins.[32]

Despite the difficulty of looking, and truly seeing this situation face-on, there is no doubt—because of Madley—that the official, now well-documented *expansive genocide* of Native Americans took place in California from 1846 to 1873. He also illustrates historically how the elected California officials were the "primary architects" of this destruction.

These officials did so by creating a legal environment where *Native Americans* had no rights whatsoever—allowing attackers to kill, move in, and take over with total impunity. Madley describes this as "building a killing machine." But no US President chose to speak out against these genocidal massacres, while two California governors "did verbally declare that the physical destruction of all California Indians was inevitable." Madley explains.

> On January 7, 1851 Governor Burnett publically proclaimed that 'a war of extermination will continue to be waged between the races, until the Indian race becomes extinct,' adding that 'the inevitable destiny of the race is beyond the power or wisdom of man to avert.' On August 11, 1852, Senator Weller—who became California's governor in 1858—went further. He told his fellow US senators that California Indians 'will be exterminated before the onward march of the white man' explaining, 'humanity may forbid, but the *interest* of the white man demands their extinction.'[33] (emphasis included)

This killing machine, initiated by California state legislation, first methodically stripped Indians of all their human rights, including their lands, which had been previously negotiated with the US government in 18 treaties which the US Senate subsequently refused to pass. As Madley describes it: "This exclusion facilitated violence ... while eroding legal and moral barriers to kidnapping, enslaving, starving, raping, murdering and massacring California Indians." All of this was funded and facilitated by state-funded militias. So, can this be characterized as California's first effort at state policing? Sadly, the history is there for all to see—if we care to look.[34]

While few state or federal actors made any effort to stop what was essentially an "extermination policy," there were some that tried. Madley helps us remember them: some agents "pleaded for change, some soldiers protected Indians"; one Kentucky senator, John Crittenden, "spoke out against 'the slavery and oppression and murders' of California Indians." All of this leads Benjamin Madley to conclude the following: "These critiques showed that alternative policies were conceivable, proposed, and possible. However, the fact remains that an anti-Indian US Congress and southern leaders like Jefferson Davis won the day. The result was genocide."[35]

Leland Stanford, while Governor of California from 1862 through 1863, signed legislation that again supported and expanded the "California Volunteers" which carried forward one of the longer and more lethal war campaigns against California Indians, totally supported first by state and then by federal dollars. Stanford's 'colossal fortune' was also built in part on California real estate—land taken from the Native American tribes of Northern California. In 1884, Leland Stanford, with his wife, founded and endowed a school of higher learning in honor of their only son who had died of typhoid—that school is now known as Stanford University.[36]

It is interesting to note that while Madley clearly documents all of this history, neither the current online biographies of Governor Stanford, which can be viewed on the "National Governors Association" website and also on the "Governors Library" website relate to any of these genocidal actions.[37] However, Berber Jin, a journalist at the "Stanford Politics," a university student newspaper, does so in a 2020 article entitled "An Overdue Encounter with the Facts." He concludes that at least "a broader project of justice and truth-telling is in order."[38] This is just one more massive past injustice that needs state-wide and national discussion.

Just to be clear, Native Americans did first suffer devastating losses from 1769 to 1846 by the Spaniards, Russians, and Mexicans during the colonization of California's coastal region. During that period, populations fell from an estimated 310,000 to approximately 150,000. But there was worse to come under US rule. From 1846 to 1870, Madley documents that Native American populations in California plunged 80% from approximately 150,000 to 30,000. In 1880, US census takers recorded only 16,277 California Indians.[39] He also informs us that many California Indians did, indeed, die of disease. "However, the massive population decline of 1846-1873 was also the product of direct killings that were perpetrated by human actors. These killings were, in turn, tolerated and often sponsored by both state and federal authorities."[40]

Madley tells the genocidal California history in chronological fashion, year by year. All of the individualized histories are arresting, and too many are just downright obscene in terms of murderous violence. Using the definition of "genocide," as recognized by the 1948 United Nations Genocide Convention, Madley meticulously describes what happened, how it happened, who supported it, who did the killing, how they did the killing, plus, how it was funded. As a result, *American Genocide* "carefully describes the broad societal, judicial, and political support for the genocide and how it unfolded."[41]

That sentence is a shocker, isn't it? We now know, for instance, that two former state governors supported and funded genocide initiatives with the support of the federal government. At the same time, the numerous local vigilante volunteers were identified as the main perpetrators. And similar to other regions, "the legacy of the state's genocidal past remains hidden in plain sight."

The difficult conclusions of the author's research are echoed by previous historians. In 1890, the California Indian situation was summarized by historian Hubert Howe Bancroft: "The savages were in the way; the miners and settlers were arrogant and impatient; there were no missionaries or others present with even the poor pretense of soul saving or civilizing. It was one of the last human hunts of civilization, and the basest and most brutal of them all." Several decades later in 1935, the US Indian Affairs Commissioner added: "The world's annals contain few comparable instances of swift depopulation—practically, of racial massacre—at the hands of a conquering race."[42]

Madley concludes, and his research definitively indicates, that California colonization efforts did indeed include "deliberate attempts to annihilate the California Indian." With this evidence, long held interpretations of Native American deaths—as they encountered white settlers—will now need re-examination. In particular: "Exceptionalist interpretations of US history—which suggest that the United States is fundamentally unlike other countries—lose validity when researchers compare the California experience to other genocides and place it within global frameworks." That is just a nice way of saying that we have preferred not to see the truth. Madley concludes that most of us today have never thought of these "pioneers" or "settlers" as "invaders," but in California from 1846 to 1873 that is often what they were.[43]

Branded as 'Less Than'

We can now begin to see that the concept of *negative difference* is deeply embedded in our society, and even at this point in time, it seems to be judged by many to be both necessary and at the same time innocuous. But every so often someone sees it for what it is. In the earlier discussed notebook, which author John Edgar Wideman kept on the 1955 trial of the two men who murdered Emmet Till—both found 'not guilty'—the author captures the essence of 'negative difference' and its difficult nuanced reality.[44]

In this notebook, the last newspaper excerpt Wideman tells us about is from an African American novelist, Chester Himes. Evidently, Himes "chose not to reside in his segregated country and probably sent his letter to the *New York Post* from Paris France." In his letter, Himes places his finger on a particular interaction that we rarely allow ourselves to approach, but we definitely need to see and understand. His letter reads:

> The real horror comes when your dead brain must face the fact that we as a nation don't want it to stop .... So let us take the burden of all of this guilt from these two pitiful crackers. They are but the guns we hired.[45]

All of that was then; this is now. Recently, part of this sad, tragic story seemed to be finally culminating. On February 26, 2020, 65 years after Emmet Till's murder, the US House of Representatives passed legislation named the "Emmett Till Antilynching Bill"—once again attempting to make *lynching* a federal crime. While many states have passed anti-lynching bills, more than 200 anti-lynching bills failed in the US Congress during the last half of the 20th century. During that period, the two bills that did pass the House were stopped in the Senate by the powerful coterie of Senators representing Southern states. But this present 21st century Bill passed 410 favoring, to 4 against in the House. However, once again, this House Bill was held in the Senate in June 2020 by the filibuster rules exercised by Senator Rand Paul, who 'has questions.' Finally, in 2022, Paul removed his objections, and the Bill was passed into law.[46] But Till's incomprehensible story continued, as does ours.

Chester Himes, in the above quote, clearly asks us to recognize and examine that murderous and antagonistic definition of *difference* and *indifference* within our society. One that engendered a classification of 'inferior' by our ancestors, and one that is currently maintained by us—all of us collectively—with or without our awareness. It does bring to mind the acclaimed philosopher, Achille Mbembe, when he contends that blackness is equated with the nonhuman in order to uphold forms of oppression. Mbembe argues "that this equation of Blackness with the nonhuman will serve as the template for all new forms of exclusion."[47]

So, what were Rand Paul's "questions" about the anti-lynching Bill? Did he want the guilt to remain specifically on those 'two pitiful crackers?' Did he want to say to all of us: 'Hey, murder is murder; it's terrible but it happens. But there is no need for us to recognize it as *lynching*. It was just an individual murder—the rest of us are innocent.' Or, maybe he just wanted us all to remain blind to the realities of our society? The noted lawyer and Harvard law professor, Derrick Bell, suggested that and more when, after years of believing in and fighting for civil rights in the US Courts, he finally decided that racism was a permanent foundation of the United States.[48] So, did Senator Rand Paul want to reinforce that societal blindness and make sure that racism remained as a foundation of our society? Of course, Rand voiced none of these ideas. But I do wonder; he had to have some reason.

A review by the Minneapolis Star Tribune of Wideman's book on Louis Till reads as follows: "Reading *Writing to Save a Life*" is to ride shotgun in [Wideman's] tricked out time machine to a familiar destination: the jagged fault lines of America's racial divide." So, all of us together—black, white, brown, yellow, and red—can do no less than keep going to see where these fault lines lead, because we absolutely do want these obscenities to stop. We want to be able to finally say that our society is no longer either overtly or subliminally racist. We want to understand how it is that our society became complicit so quietly and unobtrusively that the majority of us still do not recognize the true reality of it. Most of all, we want to learn how to change it. Understanding our history as people where many believed they were white and therefore superior, but now begin to understand we are all in this together, is one of the best ways we can go about it.

Notes

1 James M. Washington, *A Testament of Hope: The Essential Writing and Speeches of Martin Luther King*: "Where Do We Go from Here: Chaos or Community" (New York: Harper One, 1986), 562.
2 Toni Morrison, *Playing In the Dark: Whiteness and the Literary Imagination* (Cambridge: Harvard University Press, 1992), 47.

3 Morrison, *Playing in the Dark*, 233.
4 Pew Research Center, "Income Inequality in the US is Rising Most Rapidly Among Asians", July 2018, https://www.pewresearch.org
5 USA Facts, "Wealth Inequality Across Race: What Does the Data Show?" Feb. 2023, https://www.usafacts.org
6 Pew Research Center, "How the American Middle Class Has Changed Over the Past Five Decades", 2021, https://www.pewresearch.org
7 Greg Rosalsky, 6/2/2020, "Minneapolis Near the Bottom for Racial Equality", NPR *Planet Money*, July 2, 2020, https://www.npr.org
8 Department of Justice, "Justice Department find Civil Rights Violations", June 16, 2023, www.justice.gov. CBS, "Minneapolis Police systematically targeted people of color", https://www.cbsnews.com -video
9 Pew Research Center, *On Views of Race and Equality Blacks and Whites are Worlds Apart*, 1–12. Pew Social Trends, https://www.pewresearch.org
10 Pew Social Trends, "On Views."
11 Richard Rothstein. *The Color of Law: A Forgotten History of How Our Government Segregated America* (New York: Liveright Publishing, Division of W.W. Norton and Co. 2017), 237.
12 Penal Code 196, California Rules of Engagement for Use of Force by Police, 1872.
13 Teri Figueroa. "New Law to Limit Police Use of Force", *San Diego Union Tribune*, Aug. 20, 2019.
14 Deepti Hajala. "Advocates: Black Cops Not Exempt from Anti-Black Policing", *NY Amsterdam News*, 2023, https://www.amsterdamnews.com
15 Sarah Schulman. "Red Lights, Blue Lines", *New York Review of Books*, Mar. 23, 2023, 33–35.
16 Anne Gray Fischer. *The Streets Belong to Us: Sex, Race, and Police Power from Segregation to Gentrification.* (University of North Carolina Press), 2022. Anna Lvovsky, *Vice Patrol: Cops, Courts, and the Struggle over Urban Gay Life Before Stonewall.* (University of Chicago Press, 2021).
17 Sarah Schulman. "Red Lights, Blue Lines", 33–35.
18 Mariame Kaba. *We Do this 'Til They Free Us: Abolitionist Organizing and Transforming Justice* (New York: Haymarket, 2021).
19 Carol Anderson. *White Rage: The Unspoken Truth of Our Racial Divide* (New York: Bloomsbury, 2017), 43.
20 Southern Poverty Law Center (SPLC) Report, "Trump's Fear Mongering Fuels Fourth Straight Year of Hate Mongering" Spring 2019.
21 SPLC Report, Spring 2019.
22 SPLC, Hate Map, 2022, https://www.splc
23 Thomas Chatterton Williams. "Crimes of Being", *New York Times Magazine*, Jan. 29, 2017, 32. References *Writing to Save a Life: The Louis Till File* by John Edgar Wideman, emphasis added.
24 Annette Gordon-Reed, "MLK: What We Lost", *New York Review of Books*, Nov. 8, 2018
25 Annette Gordon-Reed. "MLK: What We Lost", emphasis added.
26 James M. Washington. "Where Do We Go From Here", 557, emphasis added.
27 James M. Washington "Where Do We Go From Here", 562.
28 Ibram X. Kendi. *Stamped From the Beginning: The Definitive History of Racist Ideas in America (*New York: Bold Type Books, 2016).
29 Kendi, *Stamped from the Beginning*, 3.
30 At the end of his lifetime, Jefferson freed four of the children, but Sally Hemings remained enslaved.
31 Kendi, *Stamped from the Beginning*, 10.

32 Benjamin Madley. *American Genocide: The United States and the California Indian Catastrophe, 1846-1873* (New Haven: Yale University Press, 2017), 356.
33 Madley, *American Genocide*, 353.
34 Madley, *American Genocide*, 163–172.
35 Madley, *American Genocide*, 354–356.
36 Madley, *American Genocide*, 348–350.
37 National Governors Association (www.nga.org), and California Governor's Library, https://governorslibary.ca.gov, (accessed July 15, 2020).
38 Berber Jin, "An Overdue Encounter with the Facts", *'Stanford Politics' student newspaper*, Oct. 27, 2017.
39 Madley, *American Genocide*, 3.
40 Madley, *American Genocide*, 352.
41 Madley, *American Genocide*, 8.
42 Madley, *American Genocide*, 3.
43 Madley, *American Genocide*, 15.
44 John Edgar Wideman. *Writing to Save a Life: The Louis Till File* (New York: Scribner, 2016).
45 John Edgar Wideman, *The Louis Till File*, 11.
46 Deborah Barfield Berry, "House Approves 'Long Overdue' Anti-Lynching Legislation", *USA Today*, Feb. 27, 2020. https://www.usatoday.com. David Smith, "Rand Paul Stalls Bill That Would Make Lynching a Federal Crime", *The Guardian*, June 11, 2020. https://www.theguardian.com.
47 Achille Mbembe, trans. Laurent Dubois. *Critique of Black Reason* (Durham: Duke University Press, 2017), book jacket.
48 Derrick Bell. *Faces at the Bottom of the Well: The Permanence of Racism* (New York: Basic Books, 1992). 92–104.

Bibliography

Anderson, Carol. *White Rage: The Unspoken Truth of Our Racial Divide*. New York: Bloomsbury, 2017.
Barfield Berry, Deborah. "House Approves 'Long Overdue' Anti-Lynching Legislation", *USA Today*, Feb. 27, 2020. https://www.usatoday.com
Bell, Derrick. *Faces at the Bottom of the Well: The Permanence of Racism*. New York: Basic Books, 1992.
CBS. "Minneapolis Police systematically targeted people of color", July 2020. https://www.cbsnews.com-video
Chatterton Williams, Thomas. "Crimes of Being", *New York Times Magazine*, Jan. 29, 2017.
Department of Justice. "Justice Department Find Civil Rights Violations", June 16, 2023, https://www.justice.gov
Figueroa, Teri. "New Law to Limit Police Use of Force", *San Diego Union Tribune*, Aug. 20, 2019.
Gordon-Reed, Annette. "MLK: What We Lost", *The New York Review of Books*, Nov. 8, 2018.
Gray Fischer, Anne. *The Streets Belong to Us: Sex, Race, and Police Power from Segregation to Gentrification*. The city of Chapel Hill: University of North Carolina Press, 2022.
Hajala, Deepti. "Advocates: Black Cops Not Exempt from Anti-Black Policing", *NY Amsterdam News*, 2023, https://www.amsterdamnews.com
Jin, Berber. "An Overdue Encounter with the Facts", *Stanford Politics*, Oct. 27, 2017.

Kaba, Mariame. *We Do this 'Til They Free Us: Abolitionist Organizing and Transforming Justice*. New York: Haymarket, 2021.
Kendi, Ibram X. *Stamped from the Beginning: The Definitive History of Racist Ideas in America*. New York: Bold Type Books, 2016.
Lvovsky, Anna. *Vice Patrol: Cops, Courts, and the Struggle over Urban Gay Life Before Stonewall*. Chicago: University of Chicago Press, 2021.
Madley, Benjamin. *American Genocide: The United States and the California Indian Catastrophe, 1846-1873*. New Haven: Yale University Press, 2017.
Mbembe, Achille, trans. Laurent Dubois. *Critique of Black Reason*. Durham: Duke University Press, 2017.
Morrison, Toni. *Playing In the Dark: Whiteness and the Literary Imagination*. Cambridge: Harvard University Press, 1992.
Pew Research Center and Pew Social Trends. https://www.pewresearch.org
Rothstein, Richard. *The Color of Law: A Forgotten History of How Our Government Segregated America*. New York: W.W. Norton and Co. 2017.
Schulman, Sarah. "Red Lights, Blue Lines", *New York Review of Books*, Mar. 23, 2023.
Smith, David. "Rand Paul Stalls Bill That Would Make Lynching a Federal Crime", *The Guardian*, June 11, 2020, https://www.theguardian.com
Southern Poverty Law Center (SPLC) Report. "Trump's Fear Mongering Fuels Fourth Straight Year of Hate Mongering", Spring, 2019.
US Census Bureau. Household incomes and poverty, https://www.census.gov
USA Facts. "Wealth Inequality Across Race: What Does the Data Show?", Feb. 2023, https://www.usafacts.org
Washington, James M., ed., *A Testament of Hope: The Essential Writing and Speeches of Martin Luther King*. New York: Harper One, 1986.
Wideman, John Edgar. *Writing to Save a Life: The Louis Till File*. New York: Scribner, 2017.

3
INJUSTICE: THE OPPRESSION OF GENDER

Gender injustice, its strength and impact, is out there for all to see. At least we think it is—but then again, maybe some of its most important aspects are not. The first women's rights convention in the USA was held in Seneca Falls, New York, in 1848, where 68 women and 32 men—a significant mix of black and white attendees—signed a *Declaration of Rights and Sentiments*. Patterned after the nation's Declaration of Independence, it contained these words: "We hold these truths to be self-evident: that all men and women are created equal." But the issue of gender injustice had been raised long before that meeting. Almost a century earlier, Abigail Adams, writing a letter in 1776 to her husband John, reminded him: "In the new code of laws, remember the ladies and do not put such unlimited power into the hands of the husbands." John Adams responded to her, again by letter, stating, "I cannot but laugh. Depend upon it, we know better than to repeal our masculine systems."[1]

Masculine Systems

John Adams was telling the truth—then and now—the repeal of masculine systems was not, and is not, to be tolerated even today. So it is only with faint surprise that the idea of "the female voter" in 1848 was greeted with ridicule and disbelief. No surprise either that it was another 72 years before women actually obtained the vote in 1920 by passage of the Nineteenth Amendment. But because of women's increasing optimism, it was something of a surprise when, in 1991, 71 years after obtaining the vote, Clarence Thomas was named to the United States Supreme Court, despite sexual harassment charges brought by law professor Anita Hill. Although there were others willing to testify to similar harassment charges, the Senate Judiciary Committee declined to do any investigation other than interrogate Anita Hill, a professional associate of Clarence Thomas. After the televised interrogation, Thomas was voted in by Republicans and Democrats alike.

So, having learned that difficult lesson, it was only with faint surprise that 27 years after the appointment of Clarence Thomas—and 242 years after Abigail Adams reminded her husband to do the right thing—we witnessed, in 2018, the Senate Republican majority in the US Congress, uniting with one Democrat, once again decided to maintain their patriarchal "masculine systems." They did so by voting to confirm Judge Brett Kavanaugh, a conservative whose court decisions define an extremely limiting, if not belligerent, view on women's rights.

An originalist in the same mold as Supreme Court Justice Clarence Thomas, Kavanaugh started his career working for Independent Counsel Kenneth Starr on President Clinton's impeachment. Kavanaugh has a long list of extremely conservative and sometimes punitive decisions made while serving on the D.C. Court of Appeals from 2006 to 2018. Those which are particularly important and belligerent to women include limiting rulings on Abortion and Birth Control, ObamaCare, Workers Rights, Immigration, and Affirmative Action.[2]

None of these, however, could or should have precluded Kavanaugh from being selected as a legitimate nominee. The "conservative" perspective is a well-accepted and popular political orientation, and therefore, before his nomination, he was judged "extremely capable" by review panels. So, no, it was *not* these Appeals Court decisions that made him ineligible in the eyes of so many. Instead, similar to the Clarence Thomas hearings years ago, it was *truth* and *investigation*—or lack of it. Statements made by Kavanaugh in refuting sexual assault charges brought against him by Dr. Christine Blasey Ford during a Congressional hearing for his nomination to the Supreme Court were neither adequately processed nor investigated, as indicated in the transcript of the September 27, 2018 hearings.[3]

Kavanaugh's own extraordinary anger and belligerence in responding to these charges, well documented on TV news broadcasts, increased challenges and requests for further investigation, but they were again blocked by Republicans and not adequately processed or undertaken. Senator Orrin Hatch (R-Utah) went so far as to state that sexual assault so long ago, even if true, should not necessarily disqualify Kavanaugh.[4] All of this undercut trust in the US Congress and the Supreme Court across an already dividing country.

There was also an aftermath that only a few of us heard about. During and after the Congressional Hearings, a total of 83 formal ethics complaints were filed against Kavanaugh. (Two of them were, predictably, reported to be from Democratic Party organizations.) A federal panel of Court of Appeals judges, appointed by Chief Justice John Roberts, was later appointed to investigate these 83 ethics complaints. But although the panel subsequently found that "the charges were serious," they also found that they had no authority to investigate. This was because panel members ascertained that once Kavanaugh was sworn in as a Supreme Court Justice (which happened immediately after the Congressional vote), the panel itself, as members of a lower court, had no jurisdiction. In December 2018, the review was formally closed, with no investigation and no further information released concerning the charges.[5] More recently, however, beginning in the summer of 2021, pressure began building to reopen the investigation.

So, it is not surprising that the 2018 hearings turned into—for a substantial part of the country—a formidable injustice, based on patriarchy and sexism, instead of the intended bipartisan political hearing. This change of heart by the majority of the population was individually illustrated by John Paul Stevens, a well-respected and retired Supreme Court Justice. Stevens announced that while he initially believed Kavanaugh to be well qualified for selection to the Supreme Court, "his performance in the hearings caused me to change my mind."[6]

In both of these Senate Judiciary hearings, more than a quarter century apart, Professor Anita Hill in 1991 and Dr. Christine Blasey Ford in 2018 simply appealed to the US system of government for justice. They both posed similar questions: should men accused of sexual aggression and sexual assault be appointed, with the highest honors, to the Supreme Court? Instead of investigation and fact-finding in search of a decision based on justice, what they received can only be categorized as injustice based on unadulterated patriarchy. Both men won their appointments: Thomas, 52-48; and Kavanaugh, 50-48, with Senator Joe Manchin the only Democrat voting for now Justice Kavanaugh.

The well-known feminist scholar, bell hooks, self-described as black and queer, described the finale of the 1991 Clarence Thomas hearings. Her message is condemnatory and even one-sided, but it has the ring of truth observed.

> Clarence Thomas's "declaration that he was a victim of a "high-tech lynching" ... evoked this image of white males controlling black manhood, and it was most effective. In the popular imagination of white and black folks alike, he represented the black male standing up for his right to participate fully in patriarchy, in the culture of the phallus .... After all, Thomas, even in the act of attempting to sexually coerce a black female, was only acting as white men have acted with impunity in white racist society. And in choosing to marry a white female (whose image was always behind his during the hearing) he was also expressing his allegiance to white supremacist patriarchy."[7]

The Kavanaugh hearings in 2018 did not wrap sexism into racism as did the Thomas hearings, but the continued comparisons with the Clarence Thomas and Anita Hill processes gave it a certain patina of remembrance. The rant of patriarchy, however, was definitely out there for all to see and hear, full-blast. First from Kavanaugh himself; then both Senator Republican Senators Lyndsey Graham and Chuck Grassley, the Head of US Senate Judiciary Committee, who were both remarkably forthright in expressing their churlish patriarchal attitudes and powers. Senator Grassley, when asked on camera during the hearings why there were no Republican women on his side of the Senate Judiciary Committee as compared to the multiple female members on the Democratic side, shrugged and responded, "It's a lot of work—maybe they don't want to do it." The at-odds unreality and uneasy humor evident in that remark can hardly be missed by anyone, including by other Republicans.[8]

Can we learn something basic from these telling moments? To begin with, half of the Senate was willing to say to women: "women's rights don't count" or at least that their accusations against men who are part of the club do not count as much. Does this mean they believe their beloved *masculine systems*—those systems which bring them profit, power, and control—must continue to supersede everything else? Yes, I think that is the message.

It is a message that has been brutally hammered home more recently by the extreme 2022 Supreme Court's *Dobbs v Jackson Women's Health Organization* decision. Overturning *Roe v Wade* on abortion rights means that control over a woman's own body is now no longer awarded to women themselves. Instead, this most basic right of control for one's own person is once again now vested in the power and control of men and the state, as they define it. Overturning this constitutional right is, however, not the most serious aspect of this decision. Most seriously, it indicates a turning back to a time where outright domination of the planet by humans—of women by men, of blacks by whites, of poor by rich—is believed to be right and natural.

One year after the Dobbs decision in 2023, more than 22 million US women were found to live in states where there is no reliable reproductive health care, but they do include various bans on abortion. After the Supreme Court passed the Dobbes decision, state legislatures across the country immediately introduced 563 bills and provisions to restrict abortion. Thirty of these are now laws. The laws passed have at least one, and often all four, of the following provisions: (i) a near total ban on abortion with criminalizing enforcement; (ii) specific bans of abortion by medication; (iii) bounty hunter provisions against women seeking abortion; (iii) confusion, fear, and increased public funding for pregnancy crisis centers to dissuade women from accessing abortions and suggestions they may cause mental illness, cancer, or infertility. The medical journal JAMA reported that during 2023, in those

14 states with near-total abortion bans, 64,565 women suffering rape were unable to receive abortions.⁹

Now we know. Perhaps, deep down we have always known. This is no simple decade-long, or even centuries-long, fight. Clearly, it is not about whether you believe that abortion is right or wrong; in a democratic country, that should be everyone's personal decision. No, it is a plain and simple denial of women's rights and couple's rights: implicitly including LGBTQ+ with their social and physiological decision rights and including peoples' right to marry who they choose. In other words, it goes to the heart of democracy and corrodes. It goes to the heart of equality and diminishes. Most tellingly, it goes to the heart of human connection and decreases our humanity.

So, it is not just denial of a woman's right to her own bodily autonomy that this Supreme Court has imposed. It is the fact that we are now in danger of becoming a full-fledged patriarchy, with its inherent dismissal of freedom and equality for those judged to be "less-than." This long-awaited vindication of masculine systems, as Supreme Court Justice Alito makes abundantly clear, uses *originalism* and its rulings of the 1800s and before, when patriarchy was overt and accepted. Sadly, this threatens to be where we may return to once again.

However, if we look back over the past several decades, the women's movement has not focused on this domination and oppression perspective. Instead, it has pre-occupied itself with three subjects—critical issues to be sure—all premised on the necessity for equality with men, and faith in its attainability. They are: equal pay, sexual freedom, and stopping violence against women. But as a result, unbeknown to almost all of us, we may have been diverted from understanding that the underlying tyranny of it all could re-emerge.

Tinkering Around the Edges

To verify the truth of this, let's investigate the progress made in each of these three areas, starting with women's current economic status in the United States. In terms of economic status, if we stick to the well-codified current material-oriented status-of-women inquiries, a recent Pew Center study finds that overall, women in the United States make 82 cents on the dollar in 2022 when compared with men; but in *2002* it was 80 cents on the dollar—which means a two cent raise over the past 20 years. In 2022 black and Hispanic women made 70% and 60% as much as men, respectively, while Asian women and white women made 93% and 83% as much, respectively, compared to men. All female groups are also reported to be substantially poorer in terms of household wealth than their white male counterparts.

In comparison, in the 1980s and 1990s, the pay gap narrowed by 15 cents. Researchers suggest that this differential is most likely due to running head-on into structural issues. Evidently, this means that once gaps in education and expectations are improved, the obstacles of gender discrimination, as exemplified in rules and practice, still remain.[10] Other measurements also indicate this reality. The World Economic Forum states that only 7.4% or 37 of the 500 executives of major corporations are women, compared to 1 out of every 500 in 1998. At the same time, more than 75% of US corporate board members, where those types of decisions are discussed and ratified, are white men.

In terms of political power, critical measures include nine female State Governors out of 50 in 2021; 143 women (104D and 39R) women comprise 26.7% of the 2021 US House of Representatives. In the United States, even though women are 51% of the US population, they comprise only 30.3% of the approximately 42,000 state-elected officials, and only 5.5% of these are women of color. In spite of these findings, the situation is certainly comparatively better for women than 50 years ago and has significantly improved even over the last decade—but not by as much as most of us would expect. The United Nations lists the United States as an "unimpressive" 33rd among the high-income countries in terms of gender parity for public leadership.

As a result of the #MeToo movement revelations, sexual harassment is now looked upon as one of the major factors in holding back women in almost every line of employment. For example, in recent surveys, 81% of women say they have experienced sexual harassment in the workplace. In terms of sexual violence, the Center for Disease Control (CDC) reports that one out of every three women in our society has suffered some sort of sexual assault, with more than an estimated 250,000 rapes going unreported annually.

But, of course, that is not all. With the passage of the 2022 Dobbs decision, state legislatures are now developing their own anti-women laws. Comparing Iran and its morality police to South Carolina and its present crackdown on women as "a budding theocracy," Washington Post reporter Kathleen Parker has this to say: "My beloved state continues to outperform others in all of the worst ways. Now it wants to treat abortion as murder and apply penalties accordingly…This means that an abortion could be punished like any murder, with sentences at a minimum of years in prison to, conceivably, the death penalty, though the latter isn't spelled out in the bill."[11] Parker is not wrong in her assessment—absolute control is the point.

It is exactly these elements—the economic, the political, as well as the sexual—which have been identified in both the 20th and now the 21st centuries by women in the United States as the issues of greatest importance in terms of gender inequality. Given this situation, it is disconcerting to understand not only how little has been achieved but how far there is to go. Progress is described, over and over, as 'achingly slow.'

Stories and attitudes help us understand. In terms of women's efforts for economic equality, Cheryl Sandberg's still popular book, *Lean-In*, describes how to succeed as a woman in today's corporate world. It was a popular best-seller and was told in a disarming and charming way. But no matter: it is all about how to *get along* and, better yet, *compete* in our current corporate society—not how to change it or make it better. bell hooks observed that Sandberg's focus on "individual empowerment" creates a feminism that is both instrumental and competitive in nature, thus diminishing solidarity.[12]

Maureen Dowd, a New York Times columnist, saw it similarly but described it differently. Noting that Sandberg has said that she would like to be the " pom-pom girl for feminism," Dowd states that "Sandberg may mean well … but she doesn't understand the difference between a social movement and a social marketing campaign." Dowd then observed: "she says she's using marketing for the purpose of social idealism. But she's actually using social idealism for the purposes of marketing".[13]

So what about the other key concern: sexual violence, women's sexual liberation, and control of their reproductive rights? A description of the sexual life of young millennial women and their older Gen X counterparts in 2018, again by columnist Maureen Dowd, made me, at first, shake my head in total disbelief—this couldn't be. This was *not* what, from my perspective, equal rights intended. But a check-in with my younger friends who were single or had recently been single quickly disabused me of that disbelieving notion. Dowd notes what she calls a recent 'weird pattern' about sexual encounters in both fiction and real life. Women may not be particularly attracted to a guy that they have recently hooked up with, "but they go ahead and have sex anyway."

Dowd starts with the much-discussed short story about "Cat Person" who is repelled as she watches her date undress. "But the thought of what it would take to stop what she had set in motion was overwhelming." As a result, "she takes a sip of whiskey to bludgeon her resistance into submission." Similar real-life incidents are subsequently discussed by Dowd, and she then asks: "So you'd rather have bad sex with someone who doesn't appeal to you than find a way to extricate yourself? You can Lean In but you can't walk out?"

My multi-hued friends, from younger to older, agreed that this was, too often, a description of their present reality. Some of the reasons they cited agreed almost exactly with the column. Meeting online creates an identity for your date that may be far from the truth; and yes, because guys seem to get their sex education, or at least their expectations, from online porn, there is an inherent push for women to both compete and conform—one "fear is that dating apps make women interchangeable."

Joanna Coles, former editor of Cosmopolitan and author of the book *Love Rules*, states that she was nonplused when she heard similar stories from daughters of friends. Dowd wryly observes: "and it's hard to shock someone who edited Cosmo and can talk comfortably about Pokémon porn and vodka-soaked tampons." Coles gets the last word: "Good sex is a wonderful high. It's what great novels and great music are about. And it's free! But we've lost track of what a brilliant thing it is. It's so transactional now, it's bleak."[14]

Was that representative of women's economic status and sexual rights in the world a decade ago? Does it remain so today? Is that the non-academic and short but true synopsis of our recent feminist history? Not quite, but almost. It is absolutely correct to say that women and the men who support them have made strong inroads to the injustice that surrounds us. However, it is equally true that we may have left many women—our sisters, our best friends, our friends on the margin; our daughters, our granddaughters—bereft of the key ballast they need within our current society's competing, competitive, and demeaning messaging.

The blanket overall message remains the same: be charming, be accommodating, don't get angry, be better than your competition; and if you can't be prettier or smarter, perhaps consider being more compliant. The second message is: oh, go ahead; you're beautiful; flaunt it; be sexy; you are a man's equal; so have as many sexual encounters as you can imagine! Put those two messages together, and what we have is the alarming and perilous situation described in Dowd's column!

This perilous situation has now come to fruition. In February 2023, the Center for Disease Control (CDC) reported that teen girls across the United States are "engulfed in a growing wave of violence and trauma." Some of the findings include: "nearly 1 in 3 high school girls reported that they seriously considered suicide." This is up almost 60% in a decade. Just as shocking, "almost 15% of teen girls said they were forced to have sex, an increase of 27% over two years." Then there is the high level of hopelessness and sadness among these teen girls, while boys are said to translate these same feelings into "anger and aggression." At the same time, there were positives, mainly around better behavior—less sex, smoking—and the fact that schools can make huge differences.[15]

Luckily, there have always been stalwart and far-seeing women among us. Their words gain momentum because they do not simply take an oppositional gaze and blame men; instead, they hit their mark when their words are aimed at the systems, institutions, and structures permeating our current assemblage of social relations. But a key question remains: how do we actually interpret and utilize suggested strategies in our everyday lives? How do we *not* keep that "stiff upper lip" and work for the lowest wages? Do we hide the trauma of assault, whether we are a cis-gender woman or a member of LGBTQ+, in order to move on? Do we "go for the money honey"? Sometimes we might; oftentimes we might not, sometimes we must.

Gratefully, awareness rises as instability rises. We have now become particularly aware that racism often accompanies sexism, along with classism in some essentially still unexplained manner. In effect, these big three—racism, classism, and sexism—are all intimately interconnected, and all are resting quite comfortably, as each subject chapter now indicates, in the halls of power. All the while, they continue to manage an insidious domination while still remaining essentially hidden.

We need to ask ourselves why we do not further explore this intermingling—this intersection—particularly in terms of gender, race, and class, as they are so obviously bound together within masculine systems. But despite our momentary surges of event-based anger, as illustrated in the previously discussed Senate hearings, or the strong new movement for reproductive health and abortion rights emerging from the Dobbs decision, we essentially let these moments just drift by without exploration. As a result, generalized patriarchy and the "masculine systems" specific to the United States, continue as that proverbial well-oiled machine. And yes, while I sometimes admit to overstatement, we all know that this is still true: in this context, justice is a malnourished waif sitting on a street corner.

So how do we become more honest with ourselves and dive deeper? It's not as though all of this is so well hidden; it's just that it makes everyone so uncomfortable that we simply refuse to examine it. As a result, we continue to tiptoe around the edges. But we really do need to talk about what truly matters and not use euphemisms.

Once we decide that tinkering around the edges of patriarchy is neither particularly effective nor sufficient, a good place to start is with a definition of feminism that is different from the individualistic one promoted by Cheryl Sandberg and others. This more encompassing perspective is offered by bell hooks, quoted earlier. In her book, *Feminist Theory from Margin To Center*, she gives us an action-oriented definition.

> "Simply put, feminism is a movement to end sexism, sexist exploitation, and oppression." No matter their standpoint, anyone who advocates feminist politics needs to understand the work does not end with the fight for equality of opportunity within the existing patriarchal structure. We must understand that challenging and dismantling patriarchy is at the core of contemporary feminist struggle – this is essential and necessary if women and men are to be truly liberated from outmoded sexist thinking and actions.[16]

Agreeing to this definition, which includes the intent for LGBTQ justice as well, current events designed to move us back in time tell us that we can and must organize to consciously move away from the dominating "masculine systems" which remain a substantial and bedrock part of our society today. So, is it as simple or as complicated as that? Will it be women's choice—or will it be those masculine systems proffering their preferred choice? Looking at it all with a little humor sometimes helps. One of my favorites: Girl: I want to wait; Boy: But it's so much harder for me to wait; Girl: You, as a male, should like that I'm making you wait; Boy: Why would that be? Girl: Males invented virginity.[17]

Hidden Histories

Despite the difficulties; given the research, analysis, action, discussion, conversation, introspection, and laughter—because of all of these intersections over multiple decades and several centuries—a *feminist consciousness* has developed among us. It is continuing to develop, and much has been achieved. As a result, women as a group are now in the midst of developing an "alternate future". Gerda Lerner, one of the all-time greats of feminist thought, along with bell hooks, tells us that historically, the development of the feminist consciousness takes place in four distinct stages: "(i) the awareness of a wrong; (ii) the development of a sense of sisterhood; (iii) the autonomous definition by women of their goals and strategies for changing their condition; and (iv) the development of an alternate vision of the future."[18]

At this point in time, we as women, with our accompanying male compatriots and gender-fluid comrades, have gone through the first three stages with a *reformist* perspective of achieving "equality with men." We initially thought that would be sufficient. But now, looking around us, having entered the third decade of the 21st century, we have come to realize that this goal alone is no longer sufficient, if it ever was. Lerner states it correctly.

> The system of patriarchy is a historic construct; it has a beginning; it will have an end. Its time seems to have nearly run its course—it no longer serves the needs of men or women and in its inextricable linkage to militarism, hierarchy, and racism it threatens the very existence of life on earth.[19]

But today, masculine systems and patriarchy are words that still describe the *hideouts* for those men that use sexual aggression—from words to abuse, to assault, and sometimes all the way to rape. Let me be very clear: we are not talking about *all* men, or even the majority of men. Certainly, there has been a remarkable re-orientation of women and men's relationships in the past fifty years, beginning in the 1960s. But that change in itself is insufficient. It still remains necessary to understand the dynamics and systemic structure of these self-serving, complacent attitudes and societal structures which continue to give refuge and hideouts. Ranging from the mildly stupid to the obscenely violent, we must begin to understand what it means that they are still so very much with us. In other words, if we are to truly build this *alternate* vision of the future, we need to address, upfront and loudly, new perspectives for male/female/LGBTQ+ interaction, taking away the licenses long ago awarded to masculine systems and patriarchy.

But before we can do that, we women, particularly we white women, have some other "hidden histories" to confront if we want a solid base from which to confront the wrongs of masculine systems. These wrongs include not only the economic and political inequalities but also the brutal and inhumane histories of rape and sexual assault of black women, hidden in a history in which our white foremothers were too often complicit. This brutality and inhumanity continues in our present day under the guise of female sex-trafficking for the poor and unprotected of all colors. To truly change this, we have to come clean and be truly honest with ourselves.

Yes, all women are potential victims of sexual assault and rape, but there is a critical difference among us which has not been confronted. It is truly difficult to contemplate, but it is there, and it is real. The reality is the wanton rape that black women suffered during slavery and the Jim Crow period. There has been still no real acknowledgement of it, nor has there been any investigation into old, hidden, but still ongoing legal provisions which illustrate this history's brutality.[20] Without this acknowledgement, however, there will be little chance for an alternate and more just future.

Luckily, every so often someone comes along who decides to be so strikingly and beautifully honest that we are left breathless with the gift of her courage. It was during the summer of 2020 that this gift about our "hidden histories" came about. And the timing was perfect. *Caroline Randall Williams*, a young black poet, decided to contest those who made excuses for the Confederacy statues they believed should remain in place. Her poetry has literally wiped all of the so-called excuses off the table. And in so doing, I believe she made space for all of us to rethink where we are now and how we can go forward. It would be wrong for me to attempt to summarize Randall Williams's message; instead, I put together a few of her phrases.

> I have rape colored skin. My light-brown-blackness is a living testament to the rules, the practices, the causes of the Old South.
>
> If there are those who want to remember the legacy of the Confederacy, if they want monuments, well, then, my body is a monument. My skin is a monument ....
>
> I am a black, Southern woman, and of my immediate white male ancestors, all of them were rapists. My very existence is a relic of slavery and Jim Crow ....
>
> The black people I come from were owned by the white people I come from....And I ask you now, who dares to tell me to celebrate them?...
>
> I don't just come from the South. I come from the Confederates. So I am not an outsider who makes these demands. I am a great-great granddaughter ....

> But here's the thing: Our ancestors don't deserve your unconditional pride ….
>
> Among the apologists for the Southern cause and for its monuments, there are those that dismiss the hardships of the past…They deny plantation rape, or explain it away ….
>
> To those people it is my privilege to say I am proof. I am proof that whatever else the South might have been, or might believe itself to be, it was and is a space whose prosperity and sense of romance and nostalgia were built upon the grievous exploitation of black life …
>
> Now is the time to e-examine your position ….[21]

Randall Williams was later invited to discuss her article on Trymaine Lee's podcast, "Into America." They first talked about different kinds of violence and the often intentional forgetting of them. Randall Williams was quite clear—the silencing of trauma "writes itself on generations." That is an interesting message, particularly at this point for women. Randall Williams claims that the action of reclaiming history with honesty actually begins to change things; similar to smiling whether you feel like it or not, because the smile itself actually changes your blood chemistry. This message is comforting, challenging, and troubling—all at the same time. It is *comforting* in that it doesn't require, in any shape or form, shaming anyone other than the guilty sexist and racist ancestors that we all may have—even though if we must shame them, we still also must claim them.

It is particularly *challenging* because it requires that today's men, no matter their status, who continue to use those hideouts for lesser but still similar crimes of sexual aggression, also be "shamed and named." Not in a *blaming* mode, but simply informed that these lesser crimes of sexual aggression—even in terms of communication and name-calling—have a line of violence directly related to rape—so no, not even these, are any longer acceptable. That sounds easy, but we all know it's not. It would require a hard-to-come-by *collective decision* among a substantial group of women to bind together and collectively say—'you can't do this anymore'—'none of you can do these things to any of us'—'these acts of sexual violence or aggression are no longer consented to, or acceded to, by any of us.'

The final aspect of this message proposed by Randall Williams is *troubling* for two very good reasons. The first aspect is the lesser, but still difficult. As women, we tend to let slide the smaller acts of male violence against us, or even the supposed norms of society. We make excuses that this is just the world we live in; or it is just natural for men to act that way—boys will be boys; it's just locker room talk; etc. In other words, we have allowed men to unilaterally decide what is permissible and what is not, according to their own self-serving "masculine systems"—and then we women go along with it.

We also make excuses, for sometimes we knowingly pursue and align ourselves, for our own self-interest, with a person known for these various types of ill-treatment.

The second reason that this problem is so troubling is the difficulty for women to forthrightly address it with the honesty required, particularly those of us who believe we are white. In the podcast, Trymaine Lee asks: "Do you think that white women are ready to have that conversation about how complicit they've been in white supremacy, and also just violence?… It's been pretty bad, but we never talk about that aspect of it." Randall Williams first replies with laughter: then continues, "golly you're gonna get me in trouble."

But then, once again, she is totally honest. "My answer is I sure hope so … because you know, the Ku Klux Klan would not have gotten its traction in Jim Crow without all of those nice ladies holding picnics and saying it was a community gathering, that this is a family organization. I mean white women were the backbone of the Ku Klux Klan, of establishing the cultural norms of Jim Crow. Daughters of the Confederacy erecting all of these statues … (sigh) I don't know. I'm a little scared about it. I mean if I'm very honest, I hope it'll be better. I hope that it will. I don't know."

Creating Collective Identities

That is exactly where the Women's Movement comes in today—'we never talk about it'! Yet there is so much to talk about. And it is within these participatory women's movement structures that we might. We may not have been there historically, but we are living today in a culture created within those prior norms. So, no, we can't change history, but could we together tear down the structures and institutions that live on because of that history? Once again, Caroline Randall Williams gives an indirect response but a very clear direction. She continues:

> We have to be so vigilant about knowing what our rights can be: "I get mad at the Constitution. It wasn't written for us. It wasn't even written for all white men, right?… But I love the Constitution because America as a collective identity right now, like, you know, we were raised to believe that the Constitution is for all Americans. And I am excited to find ways to make America do what its documents say instead of what its founders meant.[22]

More than 30 years ago, a legal article appeared in the University of Chicago Legal Forum entitled, "Demarginalizing the Intersection of Race and Sex: A Black Feminist Critique of Antidiscrimination Doctrine, Feminist Theory, and Antiracist Politics" by Kimberle Crenshaw. The part of Crenshaw's paper that interested me (and that I started to remember again as soon as I read Randall William's article) was the discussion of "rape as a weapon of racial terror."

Is this the reason that the word "sisterhood"—so popular in the Women's Movement during the 1960s and 1970s—is barely, if at all, heard these days? Is it because diverse groups of people have learned that it is both difficult to create and sustain without true and painful honesty? Equally relevant is the fact that today's Women's Movement, such as it is in everyday life, may have decided to perhaps settle for a quasi-consumerist type of equality rather than confront these hard truths. But that scholarly, legal article published more than 30 years ago by Kimberle Crenshaw does the confronting for us. Crenshaw begins her article with this commentary.

> One of the very few Black women's studies books is entitled *All the Women Are White; All the Blacks Are Men, But Some of Us are Brave.* I have chosen this title as a point of departure in my efforts to develop a Black feminist criticism because it sets forth a problematic consequence of the tendency to treat race and gender as mutually exclusive categories of experience and analysis.[23]

Crenshaw's article, even after 30 years (and now of "critical race theory" fame), points a way forward consistent with the hard truths that we would rather ignore. The reality is, if women are to truly abandon masculine systems and their energizing roots of patriarchy, it will require, as Crenshaw suggests, a *"quadrupled gaze"*—women's black and white gaze together, analyzing and understanding sexism and racism again together. As a result, articles such as Crenshaw's are of extreme importance because of their intent and their documentation. They analyze sexism and racism together, thereby indicating how injustice is both amplified and made more difficult to address, and I believe we can now add LGBTQ+ to this process. Thus, the analysis itself gives groups and societies a path toward redress.

The impediments, however, must be faced. For starters, the organization and utilization of such a "quadruple gaze" require that white women analyze unsparingly their current and historical standing—and within this perspective, white women and black women have much to share. But there is more: put bluntly, white women will need to, at minimum, explore their historical complicity in keeping sexist *and* racist systems alive and functioning. This poses tough questions: have they or we maintained this complicity so as to preserve the privilege of status of protection, however unconsciously? And we cannot forget the so-called protection given to white women to shield them from the existence of LGBTQ communities, so there would be no exploration or discussion of them as people from the same families. Giving up those so-called privileged but isolating protections has always meant that each one of us could be called out and cast out.

It has always been thus, and women's history certainly documents it. We now understand that 'privileged protection' is truly based on one particular premise: the requirement to maintain *negative difference e* in complicity with the power structure on the one side; and *indifference* between oneself and the *other* who, as poor, of color, or of a different gender identity, has no protection—all in order to ensure white women's protection and belonging.

There is finally another contrasting impediment that requires recognition. As a collective group, women have to ask themselves why *they* have not come together with substantial solidarity in combating their subjugation, as have racial and, to a lesser extent, class groups. The system of patriarchy simply cannot function without the cooperation of women, so why do women continue to cooperate? Both historically and in our present situation, cooperation and complicity are clear. All types of oppression and deprivations help to maintain this system, but none of these would work without the bulwark of women's cooperation.

Certainly historically and even today—given women's lack of economic and even survival power—choices of cooperation were and are, it seems, most often rational choices. But as these survival necessities have faded away for many women, that same bulwark of silent support remains. It is true that traditionally, family and power structures have made solidarity more difficult for women to achieve among themselves—particularly so because of the factor of love involved. Black women feel the necessity to protect their men because of both love and racist *oppression;* white women feel the necessity to protect, also because of love, their men, even though many have been sexist and racist *oppressors.*

But this individualistic confrontation will not solve the situation. Instead, our continuing lack of solidarity is due to something more amorphous but also more powerful. We, as women, remain lulled into unconsciousness by the continuing ability of patriarchy to *define* and *redefine* our shared culture and society! In other words, we must shift our collective gaze from the person involved to the system involved. This embrace of collective identity is the way forward. Without this switch from individual to system, new freedoms and rights may be offered, but that "do-not-cross" line of power-defining masculine systems will remain. On the other hand, consider this. Deciding to initiate that necessary "quadruple"-gaze to deconstruct our conjoined sexist/racist injustices would truly begin a new sort and sense of true belonging.

Alternative Futures

A different but congruent way to understand and fight the constellation of sexism still surrounding our society is to recognize and do away with the marginalization that patriarchy organizes and practices. These practices still exist, but they have faded to some extent for white women of financial means. But for black women, for women of diverse color, for poor women, and for those who are gender-fluid and classified as LGBTQ+, the practice has continued unabated. And now, with the new anti-abortion rules in place in many states as of 2022, white women can also no longer consider themselves to be among the protected either. Just accepting the current masculine system as a member of the previously protected groups is no longer a guarantee.

We need to learn the basics. Gerda Lerner claims that, beginning with woman as the "first other" in small homogenous communities, the rise of patriarchy gave license to the building of sexism, racism, and classism. Her two-volume opus, *Women and History*, is an extremely useful exploration of this history. Lerner's first volume, entitled *"The Creation of Patriarchy,"* describes the foundations of patriarchy in the Western world, starting in prehistoric times and ending in the Middle Ages. She first discusses the historical evidence of societies that held women in high esteem in the Neolithic and Bronze ages. As such, Lerner provides hard but still minimal evidence of "the existence of some sort of alternate model to that of patriarchy." She concisely states: "adding this to the other evidence we have cited, we can assert that female subordination is not universal."[24]

She further explains that patriarchy's establishment was not an event, as she first supposed. Instead, it was a process that developed over a time period of 2500 hundred years from approximately 3100 to 600 BC. Evidently, it took that long for women's highly esteemed reputation to suffer the "symbolic devaluing of women in relation to the divine." Lerner also documents how these multiple positive alternatives diminished as patriarchy made its appearance. One culmination is Western society's assumption, stated as a given in Aristotelian philosophy, that "women are incomplete and damaged human beings of an entirely different order than men."[25]

Gerda Lerner's second 1993 book *The Creation of Feminist Consciousness: From the Middle Ages to Eighteen-Seventy* documents how women began to free their minds from this long-standing patriarchal thought, beginning the building of a "feminist consciousness."[26] But still, we have to ask ourselves: why has this patriarchal practice endured? Yuval Harari, in his informative book *Sapiens,* uses the descriptor—"bewildering"—to describe what he maintains is this continuing universal subjugation of women.[27] After multiple investigations, he goes on to ask, "If as being demonstrated today so clearly, the patriarchal system has been based on unfounded myths rather than biological facts, what accounts for the universality and stability of this system?"[28] There is an answer. Together, we begin to see that unchallenged subconscious thought-worlds and rarely challenged conscious mythologies tend to rest comfortably together in the halls of power.

Yearning

In our current 21st century, it is fair to say, however, that more people, with increasing insight, are beginning to recognize the suffering of "otherness."

That is a major step in the right direction. Understanding the alienation, uncertainty, and fear suffered by those who have been classified as the *other*, and informed that they do not belong—blacks, the poor, women, and LGBTQ—is a profound change within society. These same sensibilities are now, more and more, encouraging all of us to finally attempt crossing those boundaries of sexism, racism, and classism—hopefully never to return. But let's not be too sanguine about these possibilities. Certainly, diversity and pluralism are nice words. But they, in no way, describe the work still to be done. Instead, I very much like bell hook's focus on the necessity for a "repositioning"; in other words, learning "how to occupy the position of the other." It may be only this which can truly deconstruct the practice of inequality.[29]

This chapter began with bell hook's Senate Hearing assessments and her definition of feminism; it ends with her unifying explanation of where I like to think we might be heading. "Yearning" is the word, she explains, "that best describes a common psychological state shared by many of us, cutting across boundaries of race, class, gender, and sexual practice." hooks explains that this intermingling: "could be fertile ground for the construction of empathy—ties that would promote recognition of common commitments, and serve as a base for solidarity and coalition." This knowledge—created by critical voices which throw aside the "master narratives"—mutually diagnoses our ills and our suffering. In turn, this allows all of us to stand up and sing our human story. And that is what is needed to begin the recognition of our common commitments. And that is also how we organize, as expressed by that hopeful, beautiful word of *yearning*. And who better than women to begin the song?

Notes

1. Alice S. Rossi. "Letter, March 31, 1776, and April 14, 1776", *The Feminist Papers: From Adams to de Beauvoir*, cited in "The History behind the Equal Rights Amendment", by Roberta W. Francis, Chair, ERA Task Force, National Council of Women's Organization, (no date). (Columbia University Press, 1973).
2. Numerous other decisions by Kavanaugh, made from an originalist perspective while on the Circuit Court of Appeals, include Digital Privacy, Food Labeling, Environment, Taxes, Financial Regulation, and Religion and School are described in seven articles on 7/09/18 by Politico staff writers, https://politico.com.
3. Transcript by Bloomberg. "Kavanaugh Hearing", *Washington Post,* Sept. 27, 2018, https://www.bgov.com.
4. Addy Baird. "Republicans were reacting", *Think Progress*, Sept. 17, Oct. 4, 2018, https://thinkprogress.org.
5. Nina Totenberg "Federal Panel Of Judges Dismisses All 83 Ethics Complaints Against Brett Kavanaugh", Dec. 18, 2018; NPR, "Brett Kavanaugh Supreme Court Nomination", consulted May 2020. https://www.Wikipedia.org.
6. Reuters. "Kavanaugh Does Not Belong on Supreme Court Retired Justice Stevens Says", Oct. 4, 2018, https://www.reuters.com.

7 bell hooks, *Black Looks: Race and Representation* (Boston: South End Press, 1992), 81–82. Republished by Routledge, 2014.
8 Senate Judiciary Hearing, CNN, Oct. 5, 2018, https://www.cnn.com.
9 Journal of the American Medical Association, JAMA. "65,000 Rape-related Pregnancies Took Place in US States with Abortion Bans", Feb. 7 and Jan. 24, 2024, https://www.jamanetwork.com. Also reported on PBS News hour, Jan. 25, 2024.
10 Aaron Gregg and Jacob Bogage. "Progress on Closing gender pay gap has stalled over the past 20 years", *Washington Post*, Mar. 1, 2023.
11 Kathleen Parker. "South Carolina's Iran-like crackdown on women", *Washington Post*, Mar. 19, 2023, https://www.washingtonpost.com.
12 bell hooks. "Dig Deep: Beyond Lean In". *The Feminist Wire*, Oct. 28, 2013, https://www.thefeministwire.com.
13 Maureen Dowd. "Pom Pom Girl for Feminism", *New York Times*, Feb. 23, 2013, https://www.nytimes.com.
14 Maureen Dowd. "What's Lust Got to Do with It?", *New York Times*, Apr. 7, 2018, https://www.nytimes.com.
15 Donna St George. "CDC says teen girls 'engulfed' in trauma", *Washington Post*, Feb. 14, 2023, https://www.washingtonpost.com.
16 bell hooks. *Feminist Theory from Margin to Center* (New York: Routledge, 2014).
17 Lela Lee. *Angry Little Mean Girls* (New York: Abrams, 2008).
18 Gerda Lerner. *The Creation of Patriarchy* (New York: Oxford, 1986), 242.
19 Lerner. *Patriarchy*, 229.
20 Derrick Bell, *Faces at the Bottom of the Well: The Permanence of Racism* (New York: Basic Books, 1992), 97–107.
21 Caroline Randall Williams, "You Want a Confederate Monument? My Body is a Confederate Monument", *New York Times*, June 26, 2020.
22 Trymaine Lee. Into America podcast: "My Body is a Monument", transcript, July 2, 2020, https://www.msnbc.com.
23 Kimberle Crenshaw, "Demarginalizing the Intersection of Race and Sex: A Black Feminist Critique of Antidiscrimination Doctrine, Feminist Theory, and Antiracist Politics" (University of Chicago, 1989, Issue 1, Art. 8).
24 Lerner, *Patriarchy*, 35
25 Lerner, *Patriarchy*, 10
26 Gerda Lerner: *The Creation of Feminist Consciousness* (England: Oxford University Press, 1986 and 1993).
27 Yuval Harari, *Sapiens: A Brief History of Humankind* (New York: Harper Collins, 2015), 145.
28 Yuval Noah Harari, *Sapiens*. 159.
29 bell hooks, *Black Looks: Race and Representation* (New York: Routledge, 1992), 177.

Bibliography

Baird, Addy. "Republicans were reacting", *Think Progress*, Sept. 17, Oct. 4, 2018, https://thinkprogress.org

Bell, Derrick. *Faces at the Bottom of the Well: The Permanence of Racism*. New York: Basic Books, 1992.

Bloomberg. "Kavanaugh Hearing", *Washington Post*, Sept. 27, 2018, https://www.bgov.com

Crenshaw, Kimberle. "Demarginalizing the Intersection of Race and Sex: A Black Feminist Critique of Antidiscrimination Doctrine, Feminist Theory, and Antiracist Politics, University of Chicago, 1989, Issue 1, Art. 8.

Dowd, Maureen. "Pom-Pom Girl for Feminism", *New York Times*, Feb. 23, 2013, https://www.nytimes.com

Dowd, Maureen. "What's Lust Got to Do with It?", *New York Times*, April 7, 2018, https://www.nytimes.com

Gregg, Aaron and Jacob Bogage. "Progress on Closing gender pay gap has stalled over the past 20 years", *Washington Post,* Mar. 1, 2023. https://washingtonpost.org

Harari, Yuval. *Sapiens: A Brief History of Humankind*. New York: Harper Collins, 2015.

hooks, bell. *Black Looks: Race and Representation*. New York: Routledge, 1992.

hooks, bell. "Dig Deep: Beyond Lean In": *The Feminist Wire*, Oct. 28, 2013, https://www.thefeministwire.com

hooks, bell. *Feminist Theory from Margin to Center*. New York: Routledge, 2014.

Journal of the American Medical Association, JAMA. "65,000 Rape-related Pregnancies Took Place in US States with Abortion Bans", Feb.7 and Jan. 24, 2024, https://www.jamanetwork

Lee, Lela. *Angry Little Mean Girls*. New York: Abrams, 2008.

Lee, Trymaine. "My Body is a Monument", *Into America podcast*: transcript, July 2, 2020, https://www.msnbc.com

Lerner, Gerda. *The Creation of Patriarchy*. New York: Oxford, 1986.

Lerner, Gerda. *The Creation of Feminist Consciousness*. UK: Oxford University Press, 1986 and 1993.

Parker, Kathleen. "South Carolina's Iran-like Crackdown on Women", *Washington Post*, Mar. 19, 2023, https://www.washingtonpost.com

Randall Williams, Caroline. "You Want a Confederate Monument? My Body is a Confederate Monument", *New York Times*, June 26, 2020.

Reuters. "Kavanaugh Does Not Belong on Supreme Court Retired Justice Stevens Says", Oct. 4, 2018, https://www.reuters.com

Rossi, Alice S. "Letter, March 31, 1776, and April 14, 1776", *The Feminist Papers: From Adams to de Beauvoir*, cited in "The History behind the Equal Rights Amendment", by Roberta W. Francis, Chair, ERA Task Force, National Council of Women's Organization, (no date). (Columbia University Press, 1973).

Senate Judiciary Hearing, CNN, Oct. 5, 2018, https://www.cnn.com

St George, Donna. "CDC Says Teen Girls 'Engulfed' in Trauma", *Washington Post*, Feb. 14, 2023, https://www.washingtonpost.com

Totenberg, Nina. "Federal Panel Of Judges Dismisses All 83 Ethics Complaints Against Brett Kavanaugh", Dec. 18, 2018; NPR, "Brett Kavanaugh Supreme Court Nomination", consulted May, 2020. https://www.Wikipedia.org

4
INJUSTICE: THE INVISIBILITY OF CLASS

I always remember this experience. As a very small child, I was walking with my uncle one rainy day when I saw a man standing in the rain and shivering. I don't think I had yet truly learned what 'poor' meant, but I knew that the man was very unhappy. So, I pulled on my uncle's jacket and said something like, 'we should do something.' My uncle Harry looked at me, looked at the man, and then took me by the hand, all the while expertly guiding me away. He said, 'Let me tell you a story.'

The story is a familiar one in America. My uncle continued, "You know Paula, if we took all the money in the world away from everyone, absolutely everyone, do you know what would happen?" Well, no, I didn't have any idea what would happen. My uncle continued: "If we took all the money in the world away from everyone, and then divided it up equally so that everyone would have the same amount—exactly the same amount—what would happen in a year?" He then answered his own question: "Well, in just one year all the people who had been rich would be rich again with lots of money, and all those people who had been poor, would be poor again!" When I asked: "Why?" He shrugged, swinging my hand back and forth, laughed, and said: "That's just the way the world is."

As Americans, we seem to have taken that phrase—"that's just the way the world is"—as a maxim to squirm out of a lot of difficult and dirty places. One of the dirtiest and most difficult is our nationally endorsed understanding of *class* in America. According to this received belief, class injustice either doesn't exist or "that's just the way the world is." Neither statement could be further from the truth. Class is endemic and often brutal in the United States, but even if we see it, we blame the situation on the individuals themselves.

Explorations

The two injustices of the first ancient wrong, racism and gender, are both quite explicit.. The primary difficulty was choosing which story to tell which analysis to pursue, as there are so many available. But class is different. Comprehensive explanations are hard to find; class as a current issue or even as an "ancient wrong" remains hard to decipher. In other words, classism is implicit, and its critical aspects are well hidden. However, in my research, I did find several extraordinary books which usefully explore our history, our economics, our politics, and our culture of class.

That said, this presentation deserves some discussion. Because class is so amorphous in our everyday world, one of the best ways to see it clearly—to understand its weight and tenacity—is to view it from different perspectives. Nancy Isenberg's *White Trash: The Four Hundred Year Untold History of Class in America* and Nell Irvin Painter's *The History of White People* provide this, but from two very distinct orientations.[1] Isenberg shines a bright light on class as practiced in white America, while Irvin Painter delves into the interactive nuances of black/white class practice. Together, they explain little-explored complexities of 'why' and 'how' certain ideas and conventions, first adopted by our ancestors in the 17th century, have continued through to the 21st. It begins to explain how these ideas are still utilized and collectively accepted as legitimate formulations of social, political, and economic life.

Forgotten History

Describing a complicated story with its diverse historical patterns, *White Trash: The 400-Year Untold History of Class in America* by Nancy Isenberg begins our efforts to understand class and offers a well-researched perspective of untold realities. Praised from all sides for illustrating "how thoroughly the notion of class is woven into the national fabric," it opens for many of us an almost never-heard-before history. While Isenberg's singular and intentional focus on white classism helpfully shines an intense light on class itself, it also reveals the implications of assumed whiteness—both issues that we as a country have continued to dodge.

Most of us were raised with the foundational myths of intrepid Pilgrims sailing away from England to what was to become known as "America;" then landing at Plymouth Rock to courageously establish a place of religious liberty and freedom. Unless we became history majors in college or were unusually voracious readers, these foundational myths of liberty, freedom, and equality have been transferred, as we became adults, to our overall, slightly fuzzy understanding of our country's history.

If we were informed of facts such as: Plymouth Rock was not so designated until the late 18th century; or the word "Pilgrim" itself was not in popular use until 1794; or even that the Thanksgiving holiday did not exist until the Civil War when it was first inaugurated (but retroactively dated to 1621) for the promotion of the then struggling poultry industry; none of these discrete facts would have been sufficient to budge our basic national myth of courage and liberty. This "sketchy understanding" of our origins, as Isenberg terms it, is based on Americans' ability to forget what does not fit comfortably into our preferred national narrative.[2] But the forgotten history of 'class'—its intertwining from the beginning with slavery and inequality— is much more relevant to our situation today than most of us have understood, and a wealth of historical evidence proves it.

Isenberg digs deep into the reality and context of this early history of class— and it's not quite what we were taught. What is now known as 'America' or the United States was shaped by two keenly felt British necessities. First, the urgency, as expressed in the early 1600s by King James I of England, to rid the area around his castle and England itself—of the "waste people"—the poor, the unwanted, the unsalvageable—thereby cleaning up the countryside. The second was the driving compulsion of then British promoters to create new wealth and riches for themselves on new "unclaimed" lands. Variations of these two ideas—ridding their own island land of "rubbish," and using "waste people" as forced labor to produce wealth in the Americas—were not only presented in tandem to potential investors of that time. These same two concepts also currently expose, according to Isenberg, "what is too often ignored about American identity." In particular Isenberg asks: "how does a culture that prizes equality of opportunity explain, or indeed accommodate, its persistently marginalized people? Twenty-first century Americans need to confront this enduring conundrum."[3]

Jamestown, founded in 1607, was one of three settlements that initially hewed to these two concepts: waste people—compared to "weeds or sickly cattle grazing on a dunghill"—were exported to America in order to clean up England, and to create wealth for English adventurers and their investors. And in 1619, Africans, stolen from Angola, first defined as 'servants'— later designated as 'slaves'—arrived in the nearly destitute English colony of Jamestown. Initially reportedly traded to the colonists for food, they quickly became an even more oppressed labor force.

But Jamestown was not working as planned for those British promoters who had strong hopes of reaping big profits. By 1625, more than 80% of the original 6000 colonists had died. But John Rolfe (husband of Pocahontas in one of our favorite, partly true, partly mythic stories) decided to import the tobacco plant, and the colony slowly began to thrive. This not only saved Jamestown but also created high demand for labor—enslaved people from Africa and more indentured servants, often from Ireland—were thus imported.[4]

These waste people, these indentured people—poor whites, convicts, Irish, orphaned boys—classified as debtors without any rights, were under risk at many levels. If parents died before the end of their four-to-nine-year indenture contract, their children automatically became "collateral assets" of the "master." Debtor contracts could also be sold or transferred to heirs. Many of the debtors came to realize, as a colony designed for profit, contracts were too often unfairly extended, and as "waste people," they were accorded no rights.

As a result of these initial unequal circumstances, a small group of planters acquired land, labor, and wealth in a relatively short space of time. Isenberg states: "Class division were well entrenched. The ever-widening gap in land ownership elevated large planters into a small privileged faction. As John Smith lamented in his 1624 *"Generall Historie of Virginia ...."* "This dear bought Land with so much blood and cost, hath only made some few rich, and all the rest losers."[5] These evolving class divisions strengthened concepts of class and property similar to those in England. I have to observe that acceptance of these divisions may still subconsciously exist if our national enthrallment with the British royal family is any indicator.

From the beginning, class divisions were not only accepted but also expected and utilized for profit. Slaves were considered not as a class of inferior humans but simply as sub-human property. Male control over women and their fertility was also critical to these English colonial concepts. Imported women were treated as 'fertile commodities.' A prospective husband *bought* a wife at a cost of 150 lb of tobacco. Thus, women were often surveyed with the same farmer's eye used to assess cattle. Virginia Company records noted: "sexual satisfaction and heirs to provide for, would make slothful men into more productive colonists."[6]

This preoccupation with control over women and their fertility intensified when the Virginia Company of London began regulating, in 1662 (remember, we were a colony), children born to enslaved African women. It was at that point that slave children automatically became, for the first time, the chattel property of the slave owner. In other words, they became "movable property like cattle. (The word chattel comes from the word for cattle.)" But even for the supposedly free, but still vulnerable, white woman, her ability as a "breeder" was paramount. At the opening of the century of settlement, English philosopher Francis Bacon noted in 1605 that wives were for "generation, fruit, and comfort".... "The act of propagation and issue encompassed children as much as calves, alike valued as the generation of good stock. Women and land were for the use of and benefit of man."[7]

"Rebellious wives" did accrue, every so often, a certain amount of standing. In the infamous Bacon Rebellion of 1676, led by then recent immigrant Nathaniel Bacon, the sitting Governor finally refused to assist settlers on the frontier because of his fear of class warfare. He was clear: "the 'Poor Endebted, Discontented and Armed' would ... use the opportunity to 'plunder the Country' and seize the property of the elite planters." When the men and their wives were brought to court, the wives of the rebellion-makers pleaded their husbands' case, and both husbands and wives were pardoned—not by the Governor, but by the King's Court. The pardons were not given out of any sympathy for the women but simply to keep profiteering peace in the colony.[8]

Years later, an English playwright, Aphra Behn, wrote a comedy based on the rebellion entitled *The Widow Ranter*, wherein one "low-born, promiscuous, cross-dressing, tobacco-smoking widow" tells a colony newcomer: "We rich Widdows are the best Commodity this Country affords."[9] Evidently, women, even then, did "know their place" in more ways than one.

Even though we know the supposed history of the Mayflower and the Bay Colony almost by heart, there was a third set of American states of which we know little. They were, however, even more fundamental to the establishment of our forgotten and often denied class history, with its enslaved underpinnings acting as a foundation. In 1663, King Charles II of England proclaimed eight men to be the "absolute Lords and proprietors" of a new colonial charter named "Carolina." John Locke, the renowned Enlightenment philosopher and later the author of several well-regarded liberty treatises underwriting the American Revolutionary War, was one of these eight men. However, he first authored the less well-known *Fundamental Constitutions of the Carolina*. This Constitution shockingly states: "every Freeman in Carolina shall have ABSOLUTE POWER AND AUTHORITY over his negro slaves" (capitalization included in original).[10]

This is certainly not the John Locke that has been presented to us in our history and philosophy books; and it says something about either the truthfulness, or the mindset, of our historians and philosophers. This contrary perspective is, however, historically accurate. Locke was, for example, a founding member and large stockholder of the Royal African Company, which monopolized the British slave trade. Thus, he was a major player in the establishment of both slavery and class, not just for the original "Carolina"—later divided into North Carolina, South Carolina, and Georgia—but also for the other colonies, and by extension, what became the United States.

Locke did much more than enhance the rule of slavery. According to Isenberg, the planters, following Locke's plans, "fashioned a ruthless class order" that melded class and slavery together. In what became South Carolina, he proposed and promoted a "semifeudalistic" and "aristocratic society" within which highly desired land rights would be allocated and controlled by the wealthy planter class. As a result, as Isenberg informs us, South Carolina rulers structured for themselves, just as John Locke had planned, a "full-blown and incestuous class hierarchy," fully embracing both the obscene institution of slavery and the vicious hierarchy of classism.[11]

Together, white planters and merchants formed a community playing to the dynastic pretensions of its small ruling class. Termed a "self-satisfied oligarchy" by Isenberg, they consciously and meanly maintained a hyper-cruel slave state. By 1700, half of the population of South Carolina was slaves, expanding to 72% by 1740. With Georgia finally adopting slavery after first attempting to be a "free labor" entity, this scourge fast became a foundational institution of the South. And by 1770, in the colony of Virginia, a bit less than 10% of white men owned over half of its land. So, for the waste people—first despised in England and now despised in the American colonies—nothing had really changed.[12] Isenberg asks a critical question:

> Can we handle the truth? In the early days of settlement, in the profit-driven minds of well-connected men in charge of a few prominent joint-stock companies, America was conceived of in paradoxical terms: at once a land of fertility and possibility and a place of outstanding wastes, "ranke" and weedy backwaters, dank and sorry swamps. Here was England's opportunity to thin out its prisons and siphon off thousands; here was an outlet for the unwanted, a way to remove vagrants and beggars, to be rid of London's eyesore population …. The investment was not in people …. The colonists were meant to find gold, and to line the pockets of the investor class back in England. The people sent to accomplish this task were by definition expendable.[13]

In these early days, there was absolutely no effort to hide the idea of class, slavery, or sexism—the existence, as it was understood, of the superior vs. the inferior, damaged, or subhuman people. Nor was there reason to hide who benefitted from this work force. It was the well-connected men with profit-driven minds; the religious leaders pontificating the necessity for those of inferior status to serve their betters; and the plantation owners establishing a "ruthless class order" based on slavery and waste people—all this served their own pretensions of "pretend royalty." It also benefitted the enslavers—and let's not forget the trading companies importing women as breeders. All these men accepted their own intrepid superiority and the other's inborn inferiority. There was absolutely nothing to hide in the early days of our country's history as a colony.

Even George Washington, for example, who served as General during the Revolution, believed that only "the lower class of people" should serve as foot soldiers, thereby sanctioning roundups of vagabonds and others to serve as needed.[14] Thomas Jefferson, author of those esteemed and inspiring words of liberty and equality in the Declaration of Independence, was also a firm believer in Anglo-Saxon superiority. Thus, he was most interested in "moderating the extremes" of class rather than promoting any sort of equality. While he pressed for some level of public education, his proposed policy was the following: "twenty of the best geniuses, will be raked from the rubbish annually, and will be instructed, at the public expense."[15]

Abigail Adams, who stood up for women during the writing of the Constitution, expressed no sympathy for the lower classes, particularly if they attempted to protest their conditions. "Ferment and commotions," she curtly observed in a letter to Jefferson, had brought forth an "abundance of rubbish."[16] John Adams, holding no belief in equality himself, believed there was a considerable amount of pretension among his colleagues in their statements and efforts pertaining to equality. He therefore chided the Jeffersonians for being publically egalitarian in their rejection of titles while individually having no intention whatsoever of "disturbing private forms of authority: the subordinate positions of wives, children, servants, and slaves."[17]

Leaders of the Constitutional Era only indirectly, if at all, acknowledged the immoral reality of slavery. And certainly, the reality of class in a country claiming equality was not acknowledged, even though it was quite evident. Did they see the irony, the immorality in this? Or, for them, was *negative difference and indifference* simply the natural order of the world? If that was indeed the case, what changed? Why, when, and how did 'class' become a "disappearing act" in our country's history, and a free and equal democracy become the apparent order, or at least the mythic order of the day? The answer is two-part: the first concerns itself with the construction of whiteness embedded not only in politics and economics but also in culture. The second part of the answer is found with the pretend rights of "the common man." The interaction of racism and class, as Nell Irvin Painter explains it, begins to answer some of these questions.

Race-Talk Heritage

Vibrant streams of white-on-white racism, of which we hear little, played out in America during the 19th and early 20th centuries, and they shape our culture and politics to this day. *The History of White People,* by Nell Irvin Painter, a Professor of History at Princeton, immerses us in the plethora of these white racist ideas about the supposed white races, entertained over the past 400 plus years here in America, and solidified by a 2000-year ancient history.

Like Isenberg, Painter describes historical patterns and a complicated story. She describes how, during the 1800s, in the midst of extraordinary change, "Race talk"—the question of hereditary descent among white people—took over the new democracy. Painter guides us through this little-known history by defining the *four enlargements* of whiteness in American society. While clearly explaining the enduring racism in the United States, Painter's book also specifies, sometimes explicitly and sometimes implicitly, the accompanying evolving nature of classism, illustrating both its opaqueness and its depth as a subject. As a result, *The History of White People* explains why and how covert systems of domination continue to not only exist but also prevail for black *and*, to a lesser extent, certain groups of white people.[18]

I have to admit, when I first became aware of this race-talk history, I characterized it to myself as just 'crazy.' But after I thought about it for a while, I realized it wasn't crazy at all. My first consideration went to my Irish ancestors and the stories I had been told. I'd grown up hearing about my great-grandfather's role as an outlawed hedge-row teacher of the Celtic language in Ireland. He and others hid in the planted hedge-rows along the country lanes in order to hold after-school Gaelic language classes because the English colonizers would not allow the language of their ancient Irish heritage to be spoken or taught in the state schools.

There was also my great-grandfather's reputed role—as our family believed—as an initiator of the late 1800s of Sinn Fein, the famed Irish Resistance. My great-grandfather would never have allowed himself to be enslaved by the English—of that we have evidence—despite the fact that the English colonizers had evidently been doing a fairly good job of that in Ireland since the 16th century, when they first invaded. As a result, the English of the 1800s era easily solved any potential problem that my great-grandfather or others may have presented by forcefully shipping him—along with other so-called misfits—from Ireland to England. And that is how our family eventually came to Canada and later to the United States.

So, as I considered my own family history, I realized that maybe this idea of white-on-white racism was not so rare after all. I started reading more Irish history—what an eye-popping history that is—the English had been practicing white-on-white racism for centuries. But this written history is primarily based on events, with not much explanation as to why and how these systems of domination came about. However, Painter's white-on-white racism explores both intent and outcomes.

When I first began to read the *History of White People*, I did ask myself—was this a new level of delusion on the part of the American people? Yes, and no. If you're living within this particular "thought world," it would probably seem eminently reasonable—as all thought worlds do when we're living on the inside. But still, perhaps this white-on-white racism was taken up primarily by the uneducated? No, that was not the case, either. Painter clearly describes how our most illustrious American leaders and philosophers adopted and promoted these white-on-white racist ideas, using them to bracket their own foundational African racist mythology.

Painter's presentation of these histories by definition of *the four enlargements of American whiteness* is particularly helpful. Through this exploration, she addresses how culture and ideas, paired with politics, can introduce, stabilize, and systemize domination. In this American situation, the major structure, according to Painter, is the introduction, stabilization, and institutionalization of a particular type of construct—the construct of 'whiteness'—in terms of both racism and classism and how they depend upon each other. These first two boundary expansions of whiteness, both of which took place in the 1800s, might at first glance seem to be an expansion of democracy. But instead, they served to lock in a rigid and cruel classism and racism for the excluded majority.

The next two enlargements, the first more than 80 years later in the 1940s and the second relatively soon after in the 1960s, began a reverse action and created definitive openings for equality. But those openings and that potential still remain to be fully acted upon. So, Painter and Isenberg together complete that first question necessary for our inquiry—"where are we?" and both begin to explain "how we got here."

The first enlargement of American whiteness took place in the early 1800s in the name of "universal suffrage." It was a big deal, despite the fact that it was so narrowly defined. The right-to-vote was awarded state-by-state to all white males, rich or poor, whether they were property owners or not. But this right-to-vote was not given in terms of democracy. Instead of expanding the ideas of freedom and equality, as Howard Zinn explains, the "right-to-vote" was, on the one hand, a pretense, given to suppress the ongoing and well-documented danger of armed protest by the poor against the wealthy.[19] But it also initiated a long-abiding social and political "narrowing" of national life: indentured servants, slaves, women, and free people of color all remained excluded, now officially so.

While the wealthy elite grudgingly welcomed white men, they did absolutely nothing to expand the standing of these new citizens. Instead, as noted earlier, the wealthy offered a pretend "democracy of manners" only, not a vote for freedom and equality. The government leaders of the era—those within the charmed class of the supposed true Anglo-Saxon Americans—felt no need to offer any assistance to people in poverty, particularly those classified as waste people. For example, the Federal Government's regulated land sales "kept prices high enough to weed out the lowest classes."[20]

At the same time, enormous effort went into keeping the preferred Anglo-Saxon or Teutonic mythic fable "enthroned" and believed. Thomas Jefferson was one of the main purveyors of this imagined, pure, superior "Anglo-Saxon" lineage, which formulated their foundational belief system and thought-world. But Jefferson and friends were just the beginning. Ralph Waldo Emerson, the widely recognized intellectual leader of the 19th century was a major promoter of this white-on-white race theory and race talk as well. Emerson's 1856 book, *English Traits,* which focused on his favorite Saxon-based racial theories—celebrating male virility, brutality, and beauty—remained popular well into the 20th century.

It is notable that many of Emerson's race theories were later incorporated into *Eugenics,* the pseudo-science of race and class that visited particular violence on poor women, both black and white. This once-popular credo formalized and incorporated, with some success, an intemperate and violent belief system of *inequality* and *inferiority* into the USA's education, political, and social systems that still exists today. It's worth noting that Emerson's Saxon-based racial theories were also key in, "exiling the Celtic Irish—white though they may be—from American identity". Painter further explains his wide impact: "*English Traits* expressed the views of the most prestigious intellectual in the United States, elevating its formulation into American ideology.... Towering over his age, [Emerson] spoke for an increasingly rich and powerful American ruling class. His thinking, as they say, became hegemonic."[21]

The second enlargement of American whiteness was begun by immigrant challenges brought about by the Civil War in the 1860s. Thousands upon thousands of Protestant Irish and German immigrants volunteered to serve in the Union forces. Their wish "to become constituent parts of *the* American" started the second enlargement of American whiteness.[22] Their Civil War service, they felt consecrated them as true citizens, to be second-in-line, right after *the* American Anglo-Saxon. Note, however, that this group challenged neither the nomenclature nor the definition of *the* American—they just wanted IN—they wanted to be part of that in-group—and they were willing to accept a competitive second place.

The eminent Anglo-Saxon-Americans had become willing—barely—to accept Protestant immigrants by the 1850s. But Emerson and his followers did their best to draw the line at *Irish Catholics*. Part of it was that they didn't want the expanding immigrant working class to vote, but there was also a particular virulent hatred. Start-up nativist organizations such as the "Know Nothings," trading on so-called patriotism, became extremely popular. "Catholic–hating fervor swept Know-Nothings into office during the fall elections of 1854, as over a million followers in ten thousand local councils seized control of entire state governments."[23]

Over several intense decades, violence had included nearly 100 Irish "butchered" and "burned" in Louisville, Kentucky, as well as the more widespread torching of convents, Catholic churches, and Irish residences. Catholics were taunted as "vile imposters" "liars," "villains," "cowardly cutthroats," "brutal, drunken apes," and characterized as "cynically crashing their way into American politics as white men."[24] Painter explains why, looking back at the 21st century, all of this seems so inexplicable.

> Today's Americans, bred in the ideology of skin color as racial difference, find it difficult to recognize the historical coexistence of potent American hatreds against people accepted as white, Irish Catholics. But anti-Catholicism has a long and often bloody national history, one that expressed itself in racial language and a violence that nowadays attach most readily to race-as-color bigotry, when, in fact, religious hatred arrived in Western culture much earlier, lasted much longer, and killed more people. If we fail to connect the dots between *class* and *religion,* we lose whole layers of historical meaning. Hatred of black people did not preclude hatred of other white people—those considered different and inferior—and flare-ups of deadly violence against stigmatized whites.[25] (emphasis added)

In retrospect—and I am not trying to make a joke of any of this—I now realize that my Irish ancestors—particularly my great-grandfather and his immediate family—certainly did *not* suffer as they might have. On the other hand, as I remember it now, in our family, like all families, there were always intimations of things not spoken in the adult discussions that we children listened to at family gatherings. There was one phrase—"black Irish"—that I would hear now and again. If I asked Aunt Mary, my favorite of our family's red-haired aunts, she would just laugh and say, "Oh, it's your Dad, with all of that black curly hair."

But if I asked my Uncle Donald, he would roar, "what it means is that we are OUTLAWS, we always have been and we always will be!" My mother would tell me to pay no attention—Donald, she said, "the intellect of the family," was just joking. However, my uncle was perhaps more of an intellectual (and an activist) than my family realized. Growing up in the 1950s, I always knew that Catholics and Protestants were divided, but I had been taught to be proud of my Irish ancestry and its orientation to music, poetry, and literature, so I didn't let put-downs bother me.

But Painter outlines a history that most of us know little of. Perhaps the best example is that of Thomas Carlyle, one of the most influential English essayists of the 1800s. Painter explains: "From his perch in London, Carlyle saw the Irish [the Irish-Catholic Celts that is] as people bred to be dominated." In a series of essays, he "juxtaposes Black Jamaica and White Connemara as our Black West Indies and our White Ireland." In a later series, Carlyle states, "The 'sluttishly starving' Irish remind him of shiftless emancipated Negroes in the West Indies." This book of essays was published in 1853 under the title *Occasional Discourse on the Nigger Question*. Obviously, my Uncle Donald knew what he was talking about. So, I now realize that the far-away imprints of virulent hatred do continue to linger, even if people do not often recognize them nowadays.

Painter's history of the second enlargement continued. The Know-Nothings disbanded immediately before the Civil War, and the virulent Irish Catholic hatred diminished as well. The Know-Nothings disbanded because they could not decide upon a consensus slavery decision. But it makes you think: at least half of these many-thousand Irish-hating people joined the fighting forces of the North to fight against slavery, while the rest joined the Confederacy? How does that make sense? It does, however, help make sense of the fact that, at the conclusion of the Civil War, the North continued its stigma-oriented *exclusion* of poor whites. In particular, Painter observes, "poor Irish Catholics remained a race apart—Celts."[26]

After the War, the so-called Southern aristocrats, also continuing their version of societal history, decided to practice a cruel and perverse sort of *inclusion*. In order to resist Reconstruction efforts, they began to induce and use their so-called white-trash contingent—remember those "rowdy and racist followers of the New South's high profile Democratic demagogues"—as a vicious rear guard to re-subjugate the now-free black people.[27] President Andrew Johnson essentially gave permission for this type of backtracking when he vetoed the extension of the Freedmen's Bureau & Civil Rights Act in 1866. (The Act was eventually overturned, but the damage had been done.)

Thus began more than 50 years of lynching, racial terror, and entrenchment of Jim Crow laws initiated by those so-called Southern aristocrats. The Southern crackers—a good portion of the poor-white community—had finally found a way to fend off rejection if they so wished, and belong within their white society—join their former enemies, the so-called aristocrats, and do their murderous bidding. But class and white-on-white hatred continued against poor whites. Hidden but not diminished, it joined hands with white-on-black race hatred, and spread across the land—from the South, through the Southwest, and on to California.[28]

At the same time, however, the Gilded Age was providing needed distractions. To the delight of new industrialists, new foreign immigrants kept pouring in. Although not welcomed by the majority, these newcomers were welcomed and immediately hired in the economic and political free-for-all, creating extraordinary benefits for the wealthy industrialists. Not surprisingly, the trope that government should not interfere with business—particularly on behalf of the poor—became increasingly popular.

The Not-So-Magical Whitening of America

The third enlargement of American whiteness was a long time in coming. It finally appeared during the 1940 war years—almost 80 years after the 1860s second enlargement. Tendrils of the enlargement began during the 1930s Depression era, as the hold of racism and classism slightly diminished. Because so many people and families became so poor so fast during the Depression, it became harder and more self-defeating to believe some of those old tropes, particularly—"poverty was the fault of the individual alone."

But this movement actually started much earlier. Painter identifies Dr. Franz Boas, an immigrant Jewish intellectual from Germany—given tenure at New York's Columbia University in 1899—as the leader of a new focus on *equality* and *inclusion*. Boas strongly questioned the accepted notions of racial and civilizational superiority. Thirty years later, he and his small team (of now well-known notables) using scientific inquiry as their pre-eminent tool, had begun to scuttle the prejudicial notions of racism—essentially wiping away the previous "blather on race." Boas first became known for his orientation toward racial equality in 1906 by accepting an invitation to deliver the commencement address at Atlanta University, a black university. Painter notes: "It was dramatic to hear Boas speak warmly across the black/white color line in an era of naked racial antagonism."[29]

But it was the 1938–1939 radio broadcast series, featuring Paul Robeson, entitled *Americans All, Immigrants All,* a title taken from a speech given in 1938 by President Roosevelt, that marked the turning point for the roll-out of the third enlargement of whiteness. Robeson, a black lawyer and activist well known for his beautiful opera singing, sang the debut of *Ballad for Americans.* In the eleven-minute cantata, he asks, "*Am I an American?*" He answers resoundingly: "*I'm just an Irish, Negro, Jewish, Italian, French and English, Spanish, Russian, Chinese, Polish, Scotch, Hungarian, Litvak, Swedish, Finnish, Canadian, Greek and Turk, Czech, and double Czech American. And that ain't all. I was baptized Baptist, Methodist, Congregationalist, Lutheran, Atheist, Roman Catholic, Orthodox Jewish, Presbyterian, Seventh Day Adventist, Mormon, Quaker, Christian Scientist, and a lot more.*" As a result, the beginning of a singular American identity was finally launched. Despite the ongoing segregation suffered by blacks and the interred Japanese, "the loudest notes of wartime stressed inclusion."[30]

After the War, Painter tells us: "The end of legalized segregation began to propel black people into national visibility as never before." At the same time, the "races of Europe" concept, with its white-on-white imagery, was beginning to disappear. Criteria, as always, shifted according to cultural and political need. But instead of black and white creating a new pluralism, it shifted again. The 1940s recreated itself in the 1950s, once again, in conflict rather than cooperation. This time, the varied components of white-on-white racism essentially disappeared with the growing acceptance of both white ethnics' and the so-called white-trash upper-strata movement into the middle class. As a result, white-on-white racism was left aside and redefined simply as white-on-black racism, with the multi-colored poverty-stricken lumped in. Thus, the opportunity of the 1950s for a cooperative strategy based on a "rallying call for greater political and economic equality" was once again bypassed.

The fourth enlargement of whiteness began 25 years later, as America's civil rights and black power movements pushed white America into the negation of race as a founding concept. The Civil Rights Act of 1964 and the Voting Rights Act of 1965 set the stage. But it was the Hart-Celler Immigration and Nationality Act of 1965, finally countering the focus of the 1924 Immigration Act emphasizing Nordic immigration, which set the fourth enlargement in motion. Instead, the Hart-Celler Act opened the door to Asians, Africans, and multiple others from the Western hemisphere—the "seed of demographic revolution."

The Voting Rights Act of 1965 also initiated a slow-moving revolution—but not in the area expected. No longer was the counting of the 'black race' meant to enforce those now forbidden Jim Crow laws, nor was it meant to keep unwanted whites and other 'races' from immigrating to the United States. Instead, Statistical Policy Directive No. 15, issued in 1977, simply directed the US Census to collect data according to four races—black, white, American Indian/Alaskan Native, and Asian/Pacific Islander—primarily for Civil Rights enforcement.[31]

This sounded neutral and appropriate, but over time, it brought a greater focus on multi-racial identities. Expansions of these identities by the year 2000 ultimately allowed "for 126 ethnoracial groups or, as the old-school purists would say, 63 races" to be identified. Of even more consequence, "Americans disorderly sexual habits" have also contributed. By 1990, heterogeneity clearly reigned. In American families, ½ of whites, 1/3 of blacks, 4/5 of Asians, and 19/20 of Native Americans were inter-related with someone of a different racial group.[32]

The Civil Rights and Voting Acts also contributed to other unexpected consequences. One of the more interesting was that in the 1990s, the construct of 'whiteness'—held up to assessment by academic and scientific research—became, for the first time, a truly *visible,* rather than invisible, norm. With whiteness and its privileges made explicit, it became increasingly difficult to define race as anything but a demeaning social construct. At the same time, the Obama years, with their increasing inclusion, prompted many to believe that the end of "race" in America was at hand. That was not the case.

Voting and Slumming

The real issues of exclusion and inequality, as both Isenberg and Painter explore them, still rest comfortably on the yet-to-be fully examined inter-related foundations of classism and racism. To clearly understand this perspective, we need to return to an aspect of America's early history when, in the early 1800s, citizenship and voting rights were awarded—state by state—to all adult white males. Because no further assistance other than the vote itself was offered at that time, the waste people, always described as hangers-on, had to overcome economic challenges on their own. This was quite unlike their middle- and upper-class counterparts, who had, as noted earlier, assistance offered to them. As a result, having little or no means to claim the arable land offered for sale, the waste people became instead the first of a continuing wave of what came to be known as the 'frontier people.'

These waste people, as landless migrants, were first renamed squatters, and their refuge became the scrubby, unwanted backwoods. However, because the political reality had changed and waste people, specifically men, were now legal citizens with the right to vote, they could no longer be viewed as a species apart. Thus, the national story had to change too. A different story, an improved American mythology, was demanded. As a result, a new and original American mythic character—the backwoods man in his cabin—emerged in the early 1800s. It portrayed a slightly cleaned-up version of a squatter, but still with a split personality story.[33]

On one side of the cracker/backwoodsman was the "homespun philosopher"—strong, brave, and independent. On the other side, the squatter remained the "ruthless brawler," "great boaster," all part of a "lawless set of rascals" as he had always been, and most importantly, always living on land he did not own. This rogue backwoodsman, with his boaster, rascal mentality, quickly became emblematic of the very real and still very overt *class distinctions* that have always been apparent since colony colonial days. However, with the rise of Andrew Jackson and the then segregationist Democratic Party of the 1820s, their political importance heightened in post-colony America. This was not something entirely new. It had been recognized from the early 1800s that the "backwoodsman/squatter" would have to be wooed for his vote. As Isenberg explains, "the ubiquity of squatters across the United States turned them into a powerful political trope."[34]

Thus began the trade: instead of property, freedom, and equality, the squatter—now renamed as the mythic frontier man—was offered that pretend *"democracy of manners"* only. It was to be their right to boast and act out, supposedly accepted and in sync with their white, wealthy masters for just a moment—*all in trade for their vote*. In other words, what the moneyed and property-owning class did was create a deceptive act of cheap hypocritical acceptance. It was a hypocritical offer that was then accepted, although with some resentment, by the backwoodsman/squatter. As a result, *identity politics*, featuring cover-ups and conspiracies, was introduced for the first time. So, this new mythic character—the 'squatter-frontier man'—never became *the common man* as he could have, a rallying call for greater political and economic equality.[35]

So why did it work? There is both a simple individual answer and a more complex class answer as well. Simple: the backwoodsman himself believed—with resentment—that he was not only of a lower class but also of an inferior white race. On the other hand, from his perspective, at least he was white. The leaders—politicians and presidents—agreed. But they also were individually adamant in their own superiority beliefs of white-on-white races and because they were the superior race, knew they deserved to reign as upper-class rulers. As a result, these elite whites were not about to allow backwoods waste people—the same people who had, not too long ago, belonged to them as their actual property, or at least as their renters—to potentially join them at their level.

The wealthy would, however, trade votes for a more complex set of class reasons. As the superior white race, they believed they could manipulate the situation by letting the backwoodsmen have their short-lived fun while at the same time garnering support for the issues that they, the wealthy elite, felt were appropriate. This was the essence of cheap vote-getting. And it was worth the small inconvenience for the immense political power that the squatter vote would provide them. But, most important, it also kept the class divisions intact. The wealthy would never allow the poor to become their equal, challenge their superior status, permit substantial property ownership or hold public office. No, that would absolutely not happen.

It is interesting. The American white upper class was clearly on to something and was not alone in its perceptions. Thomas Piketty, in *Capital and Ideology,* systematically describes a similar process, in all of its varieties, which has sustained inequality over a millennium around the world. Like others around the world, the American wealthy evidently intuitively understood that even a simple act of "cheap hypocritical acceptance" was extraordinarily important to the lower classes, in their case, the waste/squatter/ backwoods people. Their acceptance, no matter how hypocritically offered, gave waste people an *identity* that associated them with the white rulers of America. Most importantly, this identity definitively disassociated the waste people from the enslaved black people who, like them, were poor and struggling.[36] Identity politics were at work for the first time.

Piketty's research, illustrated in multiple cultures and nations, clearly specifies that this type of "identitarian" membership, no matter how falsely conveyed, "mutes" attention to *economic* and *inequality* disparities. As a result, in America, the wealthy were allowed to essentially hide, or at least gloss over, class as unimportant and minimal. Most importantly, this "hypocritical acceptance"—using the democracy of manners—divided poor whites and blacks. The wealthy thereby avoided the potential economic wars if workers aligned together across black and white.

And, because the white underclass identified with the wealthy, many of them became willing to do the racist bidding of the wealthy Excluders over the decades and centuries—always in trade for their continued treasured white acceptance and alliance. In this 21st century, former President Trump re-initiated these same alliances. However, it is the white working class that is now labeled racist, while those in the wealthy upper class, who also continue this practice, bear no such automatic stigma.

This cycle, that began in the 1800s, continues today. Nancy Isenberg allows James Fenimore Cooper, the 19th-century novelist, to effectively sum up the reality of the situation today as then: "democratic boasting was a cheap price to pay for ensuring that real social leveling did not erode set-in-stone class divisions."[37] These same set-in stone class divisions remain with us even now, and the same semi-hidden processes continue their work—in particular, dividing white, black, and poor populations.

Painter tells the same story from her chosen perspective. She points out that there is a second and equally important reason why these class divisions have continued to work: the expanding *construct of whiteness* was an essential prop by which we mythologized our country's history as a democracy. But now, as democracy and rights have created real change since the fourth enlargement began in 1964, simple alignment of unspoken white identity is not sufficient. So, as Thomas Piketty suggests in *Capital and Ideology*, violence has joined the mix.[38]

Bamboozled

People get jittery when confronted with difficult truth so we have to sort of sneak up on it. Both Nancy Isenberg and Nell Irvin Painter help us face reality by openly and succinctly exploring the long, hidden history of class hostility, underwritten by racism. This situation, particularly when protected by hard-to-detect deceit, doesn't change easily. However, Painter and Isenberg's separate but similar conclusions offer us an important way forward. Both conclude that we as a society accept deceptive rhetoric to hide the obvious social reality of poverty, classism, and racism in a nation that supposedly values and offers both liberty and equality.

Nell Irvin Painter explores this class hostility and racism through the lens of the suburban generation and the price they pay for their 'whiteness.' She explains that "white ethnics," those whose immigrant grandparents were not the "right white," were finally becoming comfortable in their covenant - restricted suburbs, and finally feeling as though they belonged. But then everyday life unexpectedly changed. For the first time, inclusion of well-educated, middle-class blacks was now the expected middle-class norm. But sadly, these ethnics wasted no time in hurriedly and wholeheartedly rejecting any such inclusion. Instead, 'they told themselves that they had worked hard', and the 'others' had not. Evidently, they resentfully chose to forget the generous GI Bill, and the FHA support for education and housing that they had received from the Government, from which blacks had been excluded.

Nancy Isenberg also reviews the hostility perspective of white, now middle-class suburbanites similarly, but focuses instead on the impoverished redneck roots of many. This group, as "upscale rednecks," drew a clear class line between themselves and their lower-class redneck cousins, whom they now classified as white trash. Given their improved status, these upscale rednecks vociferously objected to federal aid for the poor and pushed back to Government proposed anti-poverty programs. As such, they chose to forget the experiences of their parents.

Isenberg explains. What they chose to forget is that the parents and grandparents of today's middle-class and often prosperous upscale so-called rednecks "escaped the tar-paper shack only with the help of the federal government". ... "Moving up meant staying ahead of those still trapped in the 'poverty ditch.' But rather than help others escape destitution, this new addition to the middle-class deeply resented a government that wasted money on the poor."[39] And no matter what, they could not afford any association with either the so-called black degenerate families or the still poverty-stricken whites. So, the long enduring heart of divisive animosity, created by that long-ago *democracy of manners*, continues to run amok in our society.

At first glance, this discussion seems remarkably low-key, doesn't it? But it's not. Both authors have brought us face-to-face with difficult truths that too often remain hidden. The conflation of class and racism caused by deception—with its resulting and continuing confusion—serves to hide both class and racist exploitation. This continuing deception, practiced either as perpetrator or received as a vulnerable and sometimes unaware recipient, has kept us from evolving as a true and equal democracy. And in this 2020 decade, it remains in full force. OK, that's the measured way of presenting it, but this is worth a bit of a rant.

Let's be clear. This is *not* just History 101—we need to understand that we have been BAMBOOZLED—deceived, duped, hoodwinked, and confused, over and over again, and it continues in full force today! Yes, it could be different, and we should be really mad and really tired of it by now. Painter and Isenberg actually point out what's going on over and over. It is *deception and benefit of white identity falsely offered*—both then and now—by the powerful and wealthy few to the struggling class who wish to be so associated above all else. One measure of whether we have been well and truly bamboozled is *if* the same bamboozlement has been pointed out many times before. If people told all of us way back then, and even not too long ago, that this is the way it was and is—but we didn't either pay attention or disagreed and said no—that this is not the way it is—well then?

If that is the case, welcome to bamboozlement. Novelist Toni Morrison has explained more than once how the presence of blacks enables a bonding by whites across a vast socioeconomic divide. When asked why blacks and whites can't bridge the abyss in race relations, Morrison replied: "because black people have always been used as a buffer in this country between powers to prevent a class war, to prevent other kinds of real conflagrations"[40] With that quick illustration and knowing that there are so many more, I think we might agree: We have been totally and truly BAMBOOZLED—all of us. So, it's *not* "just the way the world is." It could be different. Now we just have to figure out how to get out of this confused state.

Notes

1 Dr. Nancy Isenberg is the T. Harry William professor at Louisiana State University.
Dr. Nell Irvin Painter is Edwards Professor of American History at Princeton University, emerita.
2 Nancy Isenberg, *White Trash: The 400-Year Untold History of Class in America* (New York: Penguin Books, 2017), 7–8.
3 Isenberg, *White Trash*, 2, 3.
4 Isenberg, *White Trash*, 23–25.
5 Isenberg, *White Trash*, 26
6 Isenberg, *White Trash*, 26–28.
7 Isenberg, *White Trash*, 41.
8 Isenberg, *White Trash*, 38.
9 Isenberg, *White Trash*, 40.
10 Isenberg, *White Trash*, 43.

11 Isenberg, *White Trash*, 28, 26.
12 Isenberg, *White Trash*, 47.
13 Isenberg, *White Trash*, 10–11.
14 Isenberg, *White Trash*, 87.
15 Isenberg, *White Trash*, 85, 91.
16 Isenberg, *White Trash*, 97.
17 Isenberg, *White Trash*, 99.
18 Nell Irvin Painter, T*he History of White People* (New York: W.W. Norton, 2010).
19 Howard Zinn, *A People's History of the United States* (New York: Harper Modern Classics, 2015), Ch5.
20 Isenberg, *White Trash*, 112
21 Painter, *The History of White People*, 183.
22 Painter, *The History of White People*, 183.
23 Painter, *The History of White People*, 132–150.
24 Painter, *The History of White People*, 148.
25 Painter, *The History of White People*, 148.
26 Painter, *The History of White People*, 132.
27 Painter, *The History of White People*, 150.
28 Kevin Waite, *West of Slavery: The Southern Dream of Transcontinental Empire* (Chapel Hill: Univ of North Carolina Press, 2021). The movement of this Southern racist and classist ideology was originally believed to remain in the South. But more recent research has identified its spread across the United States.
29 Painter, *The History of White People*, 232.
30 Painter, *The History of White People*, 233.
31 Painter, *The History of White People*, 358.
32 Painter, *The History of White People*, 384.
33 Painter, *The History of White People*, 385.
34 Isenberg, *White Trash*, ch5.
35 Isenberg, *White Trash*, 112.
36 Isenberg, *White Trash*, 131.
37 Isenberg, *White Trash*, 131.
38 Thomas Piketty, *Capital and Ideology*.
39 Isenberg, *White Trash*, 277–278
40 Toni Morrison, "The Pain of Being Black", *Time,* May 22, 1989, as quoted in Derrick Bell, *Faces at the Bottom of the Well: The Permanence of Racism* (New York: Basic Books, 1992), 151–152.

Bibliography

Bell, Derrick. *Faces at the Bottom of the Well: The Permanence of Racism*. New York: Basic Books, 1992.
Irvin Painter, Nell. *The History of White People*, New York: W.W. Norton, 2010.
Isenberg, Nancy. *White Trash: The 400-Year Untold History of Class in America*. New York: Penguin Books, 2017.
Morrison, Toni. "The Pain of Being Black", *Time,* May 22, 1989
Piketty, Thomas. *Capital and Ideology*. Cambridge: Belknap Press of Harvard University, 2020.
Waite, Kevin. *West of Slavery: The Southern Dream of Transcontinental Empire*. Chapel Hill: University of North Carolina Press, 2021.
Zinn, Howard. *A People's History of the United States*. New York: Harper Modern Classics, 2015.

5
ENDLESS WAR

On my first trip outside of the United States, I visited Algeria. Entering the small rural home of people who had invited my friend and me to tea, I immediately noticed a photo of President John F. Kennedy prominently displayed in the living room. I was surprised, so I asked why. Our host replied that it was because the President and the United States stood with Algeria in support of their nation's liberation from colonialism. Decades later, I was riding in a taxi in Accra, Ghana, several weeks after the attacks of 9/11. I realized that the taxi driver was watching me in the rearview mirror as he drove. Finally, he asked, "You're an American?" When I answered yes, he abruptly went to the curb and stopped. I started to open the door and get out because I thought he wanted me to leave. Instead, he opened my door, went down on one knee, grabbed my hand, and kissed it, saying: "We are all so sorry for the suffering that your great country is enduring." That was then; this is now. Times have changed.

But the question should be asked—why have times changed so radically? Into the 1960s and for some time after, we were recognized as a country of great stature. Now, in the 21st century, we are a country that no one quite trusts. Basically, our stature was based on our ideals of freedom, equality, and democracy; we were known to stand for liberation, however imperfectly. Our overt descent, starting with the Vietnam War and its support of France as a colonial power, was a clear and public repudiation of this longstanding reputation. But in reality, it's more complicated than that. Tracing the path that leads to endless war, the loss of an international reputation, and the development of a violent national gun culture begins with untangling a complex, shape-shifting history.

Serving American Interests

Back in the 1960s, Robert McNamara, head of the Department of Defense in President Kennedy's administration and continuing in the President Lyndon Baines Johnson (LBJ) administration, made a little remarked upon but highly consequential decision. Observing that the manufacture of armaments was quite expensive with extensive cost over-runs, McNamara decided to limit armament costs. He did so with three methods—each relatively innocuous alone, but together they began the creation of a continuing platform for war. The methods are: first, increase the production of armaments; second, expand government proffered contracts to the private sector; and third, sell the weapons surplus abroad. Thus began our country's initially unremarked expansion into both the never-ending-war-syndrome as well as the mass-shooting gun culture where we now find ourselves today.[1] This is where it all begins—this is where our endless war begins.

As McNamara saw it at that time, his new policy would reduce the cost of government R&D and serve American interests. As a businessman and former president of Ford Motor Company, this commodity-sale orientation was explained in terms of the then relatively small *$3 billion* general trade deficit (in comparison, the 2020 trade deficit was over $800 billion). All of these military sales were done in the name of economy, efficiency, and business values and included support for the possibility of expanded military operations—in other words, sensible short-term reasoning. However, deaths, casualties, emotional suffering, placing us on an amoral, if not immoral, path to endless violence and war—all part of the longer-term consequences—were not included.

It was no surprise, with his decision made to sell armaments, that Robert McNamara then began to sing the praises of military sales. In a 1966 Congressional hearing, McNamara testified that DOD, with others, projected $1.5 billion worth of military export sales during the fiscal year 1968. Against mounting criticism of arms sales by members of Congress—as one senator put it: "it just smells"—McNamara defended it as "good business." He argued: "We should not kill the goose that has laid these golden eggs, and these are eggs that help us militarily, and eggs that help us commercially. We have been paid $5 billion in cash in this period [1962–1966)]."[2]

Pay attention to that phrase—"these are ... eggs that help us commercially." With that, McNamara set a new precedent: arms sales were now an international business venture. Revenues from other countries would be used to supplement DOD revenues, but no public accountability was built in or intended. And over time, it turned into a private enterprise goldmine. Remembering the $1.5 billion of sales projected in 1966, it is interesting to note that Reuters reported sales of military armaments made in the United States to foreign governments rose another 33 percent in 2018. That means the 1966 projected sales of $1.5 billion increased to *$55.6 billion* in 2018. This was an increase of more than *one billion dollars per year* over a period of 52 years.

McNamara's decision also had an essentially unknown history behind it that facilitated these military sales. However, those parts of our history giving impetus to these activities are not easy to find; they are available only if one looks for them. Nor are they part of normal American education curriculums in school, or even as part of our American mythology. However, Daniel Immerwahr's *How to Hide an Empire: A History of the Greater United States* does reveal this history, changing our collective understanding of the United States role as a superpower.[3]

Immerwahr informs us that beginning a few years after our geographical boundaries finally stretched "from sea to shining sea," the United States began to annex overseas territory. Between 1857 and 1902, the United States acquired 94 uninhabited small dots on the world map, including a few populated islands in the Caribbean and Alaska. Did you know that? I didn't. This expansion policy continued with the winning of the Spanish-American War in 1898, ending Spain's control of her colonies and assuring the United States of a new predominance in the Caribbean and the Far East. These islands included Puerto Rico, the Philippines, Guam, and Hawaii as well; later, the United States purchased the Virgin Islands from their colonial rulers in 1917.

Once the earlier debates concerning territorial usage were forgotten by US citizens, the initial rules and agreements could be changed with little or no public oversight. So, while members of Congress were forthright in their wish to deny citizenship to those whom they referred to as "inferior peoples," they also recognized the territories were useful to both the national economy and the US military establishment. For example, efforts to *standardize* manufacturing around World War II—its outputs as well as its processes—played an essential role in the gaining of influence for the United States in the 1940s and 1950s. In particular, bullets, rifles, and war gear of all types were standardized with many manufactured in these territories, initiating the perfect market-assist for Robert McNamara's militarized business platforms.[4]

By the 1960s, military interests and increasingly consumer-oriented business interests continued to spur these expansionist acquisitions. For example, the Northern Mariana Islands—acquired after World War II as a "territory"—have been, since 1975, a part of the US Commonwealth, which particularly illustrates how these territory expansions have benefited US business profits. Because of US Court decisions over time, these territories are subject to some US laws, but not all. For example, for purposes of trade, the islands were counted as part of the United States. On the other hand, the federal minimum wage law was waived, along with OSHA requirements for worker safety.

Immerwahr reports: "The combination was potent: a legal environment where foreign workers could toil for paltry wages with little oversight to stitch garments labeled MADE IN THE USA." As a result, operating at its peak, the local island garment industry sanctioned as a US product, delivered clothing worth a billion dollars wholesale to popular US retailers, including The Gap, Anne Taylor, Ralph Lauren, Alvin Klein, Liz Claiborne, Target, Walmart, and J. Crew.[5] We now realize that "national interests" does not necessarily mean "in the public interest," but too often simply means "in national business interests."

Another of the more breath-taking set of facts brings into focus a procession of American military decisions and actions for business expansion, from the beginning of our country's history, that culminates in a reality of which few of us are adequately aware. Since its founding, for example, the United States has initiated more military incursions than it would like to admit, but always we were told, "in the service of national interests." How many of us knew that? I certainly didn't. The Marine Corps *Small Wars Manual*, published in 1940, for example, gives us precise numbers on these incursions: "The Marine Corps has landed troops 180 times in 37 countries from 1800 to 1934."[6] Since World War II, war historian William Blum counts a total of 26 more.[7] Landing US troops 180 times in 37 countries from 1800 to 1934 and another 27 from 1945 to 2019 totals 207 armed US incursions in 64 countries.

The good thing was, in those early days, if the US military invaded they did indeed leave. They actually did so after each of the 180 invasions made in 37 countries between 1800 and 1934. As a result of these military incursions and business expansions, however, "the Greater United States included some 135 million people living outside the mainland" by the end of World War II.[8] At the same time, multiple Senators continued to be adamant—*no statehood for brown people.*

But few on the "mainland" were aware of this larger population or of the internal legislative debates. So as a country, we did not welcome them with expected statehood. Instead, the United States continued to maintain a quasi-colonial status as a commonwealth for each, with its accompanying economic benefits to the United States. However, much to our benefit as a country, we did not physically occupy the annexed countries but only made agreements of various kinds with them. Because of this, the United States did not acquire the reputation as a colonizer. However, the United States, according to Immerwahr, did acquire "a new sort of influence."

Accountability Resistance

McNamara's decision to manufacture and sell armaments had other consequent impacts on our country. Fast forward to the present, we are now the #1 armaments manufacturer and arms merchant in the world and have been since the mid-sixties. The pace continues, irrespective of administration. According to the Stockholm International Peace Research (SIPRI), in the 2013–2017 timeframe under the Obama administration, international weapons sales increased by 10% within the United States, with weapons exports increasing by 25% during that same period.[9]

Today, our US military budget is larger than the next ten largest countries put together, including Russia and China. In 2015 *world* military spending was more than $1.6 *trillion*. US military spending, although it is only 4.4% of the world's population, accounted for more than 34% of that total. The 2024 National Defense Authorization Act (NDAA) is $883.7B. Yes, that is right, Billion dollars with a capital B. But the Department of Defense resists audits. Not surprisingly, DOD has a long history of denials for any level of auditing requested by the taxpayer or various civic organizations. But of critical importance, they also resist audits when *required by the Federal Government*. And nothing is done about it.

That's right; so far, nothing, even though Congress has the constitutional right *and* duty to initiate and track government expenditures. Evidently, DOD is allowed to do so because there is a bipartisan moderate-middle in Congress who are determined to allow DOD both absolute freedom and secrecy, although this is clearly against our constitutional mandates. But our Congressional representatives did try twice. In 1990, Congress passed the "Chief Financial Officers Act," which requires all federal government agencies to develop budgetary systems ready to be audited and to subsequently submit to annual audits. All federal agencies have been in compliance with this law for a number of years, except for the Department of Defense. Instead, the DOD has literally been "making up numbers" in its financial reports for Congress. Senator Charles Grassley, a staunch Republican representing Iowa and a critic of DOD's financial practices and its stalling, stated on the Senate Floor in September 2017 that the Department of Defense failures reflects "26 years of hardcore foot dragging," where "internal resistance to auditing the books runs deep."[10]

Too often, criticism of military budgets is portrayed as being anti-patriotic. But that is only what profiteers want us to believe. Former President Eisenhower, Army General and staunch Republican, explained it this way in his farewell address at the end of his presidency in 1961.

> Until the latest of our world conflicts, the United States had no armaments industry …. But now we can no longer risk emergency improvisation of national defense; we have been compelled to create a permanent armaments industry of vast proportions …. In the councils of government, we must guard against the acquisition of unwarranted influence, whether sought or unsought, by the military industrial complex. The potential for the disastrous rise of misplaced power exists and will persist …. We should take nothing for granted.[11]

Eisenhower goes on to say that "only an alert and knowledgeable citizenry can compel the proper meshing of the huge industrial and military machinery of defense with our peaceful methods and goals, so that security and liberty may prosper together." Have we been alert and knowledgeable enough? It doesn't seem so.

But Congress did try one more time. In early 2018, the accounting firm Ernst & Young was hired to complete the Congress-requested independent DOD audit. However, later that same year, this well-regarded accounting firm announced they were unable to complete the assignment. Ernst & Young concluded that "the DOD's financial records were so riddled with bookkeeping deficiencies, irregularities, and errors that a reliable audit was simply impossible." Deputy Secretary of Defense Patrick Shanahan replied: "We failed the audit, but we never expected to pass it … It was an audit on a $2.7 trillion organization, so the fact that we did the audit is substantial."[12] Six years later, there have been no further attempts.

Mr. Shanahan's statement essentially acknowledges that secrecy and incompetency are not only tolerated by the Congress and citizens—or at least some majority thereof—but also preferred by DOD. Because of DOD's evident "acquisition of influence" over both political parties, they believe they can continue to demand that secrecy. Anything like transparency and accountability to either Congress or the American people is evidently not considered necessary. Few people seem to care—once again, it's just politics—until it's not.

Believe it or not, it gets worse. An outside university team, working in the Defense Department's Office of the Inspector General, then spent months in 2018 scrutinizing DOD financial statements, starting with FY 2015 and going all the way back to 1998. Their objective was to figure out both the method and scale of DOD accounting. They found that the primary method is moving the one-year money received from Congress into multi-year pools of money, where the new amounts of money are referred to as 'plugs.' Shifting these plugs multiple times from their initially congressionally authorized purpose to various funds creates the necessary duplicity: "the funds become virtually untraceable."[13] One Pentagon lobbyist, who insisted on anonymity to keep his position intact, said this is a preferred and well-recognized process. Reviewing reports from 2015 back to 1998, the university team found the following:

> The amounts of money reported as having flowed into and out of the Defense Department were gargantuan, often dwarfing the amounts Congress had appropriated …. In all, *a mind-boggling $21 trillion (at a minimum) in financial transactions between 1998 and 2015 could not be traced, documented, or explained,* Skidmore concluded. To convey the vastness of that sum, $21 trillion is roughly five times more than the entire federal government spends in a year. It is greater than the US gross national product; the world's largest at an estimated $18.8 trillion. And that $21 trillion includes only the plugs that were disclosed in reports by the office of the Inspector General, which doesn't review all of Pentagon spending.[14] (emphasis added)

So, while the military and their private contractors are utilizing almost 2/3 of the discretionary spending budget—a substantial proportion of it spent on cost overruns—all other government activities have to make-do with a bit more than a third of that same budget. Do you hear appropriate questioning about this outsize imbalance? I don't.[15] Does it serve the interests of the American people? Absolutely not. Does it maintain our previous international reputation with its ideals of freedom, equality, and democracy? Again, the answer is no.

We do, however, now know, thanks to Immerwahr's research, that we have at least 800 military installations around the world.[16] If only that were all—but it's not. After the 9/11 attacks, Vice President Cheney announced that the government would have to use the "dark side." Did many of us know what that meant at that time? I don't think so, but we quickly found out—it was in all of the news. Islands, with their 'sometimes American, sometimes not' existence, were the perfect places for illegal activities and deniable CIA 'black sites.' International law prohibits torture, so, as one person put it, we exported it, hiding behind the phrase 'extraordinary rendition.' Immerwahr observes: "This is the shape of American power today. This is the world that the United States made."[17]

Eroding the Peaceable Bias

McNamara thus began a process in which he did not anticipate the outcome. It is one that moved our country away from what it was supposed to be—in our founders' eyes, in the world's eyes, and in our eyes—a peace-loving nation. As a result of the historical missteps just discussed, in this third decade of the 21st century, our current US leadership stance is no longer the *peace-oriented vision* held by the framers of the United States Constitution.[18] If we look at our history, clearly our founders went out of their way to avoid the then already well-known incentives to war. They had seen too many kings and monarchs in Europe take their people to war for self-interested and profiteering reasons. Therefore, the Constitution framers deliberately designed our government so that we could not be driven into war by an authoritarian *preemptory executive directive*. Instead, they divided our government into three separate entities, placing war-making powers not with the Executive branch, not with the Judicial branch, but very deliberately with the Congressional branch.

Congressional power, as given by the Constitution, now requires Congress to approve or disapprove any *continuing* military initiative and then requires a scale-down of military forces after the completion of a war. So, while short military incursions did not require either a build-up of military forces or Congressional approval, longer wars emphatically did require Congressional approval. Despite our supposed love of the Constitution, this peace-oriented aspect is rarely quoted or discussed. But Rachel Maddow reminds us in *Drift*, "Those worries about the inevitable incentives to war were part of what led to the division of government at the heart of our Constitution, building into the structure of our new country a *deliberate peaceable bias*" (emphasis added).[19]

Because the Founders understood that the temptations were great, they also strengthened our nation's early aversion to war with the reality of the *citizen soldier*, designed to countermand the temptations of a standing army. Of course, there were always workarounds, as illustrated by the documented short-term military invasions beginning in the 1850s. But Congress was still able to basically maintain the separation of powers defined in the Constitution—until the Vietnam War. Evidently, those multiple quick retreats from new territories, as required by our Constitution, were at least part of the reason that our country was perceived to be a true supporter of liberation for the first 200 plus years of our country's existence.

But that separation of the three branches of government was noticeably diminished and perhaps lost when the US Congress refused to declare the ongoing Vietnam military conflict as a war in 1964. Instead, Congress simply passed a Resolution authorizing "military escalation," allowing the President and the Executive branch to proceed as they wished. President Lyndon Baines Johnson did exactly that: he committed 20,000 armed troops to Vietnam in late 1964 without any Congressional authorization. This decision was based, in large part, on the assurances that Robert McNamara, as Secretary of the Navy, gave to Congress, telling them there was no American involvement in the Gulf of Tonkin attack—which was later acknowledged to be a lie on McNamara's part.[20]

Sometimes too many years get in the way, and we don't connect the dots. The Tonkin Gulf Resolution gave a new and previously unprecedented "expansion of the presidential power to wage war that is still used regularly."[21] The initial result was an erosion of trust in their government experienced by the American people as the reality of the Vietnam War hit home, starting with escalating numbers of dead on all sides. According to the US government archives, a total of 58,200 soldiers were killed in the war, and various death estimates on the Vietnamese side—including soldiers and civilians—range from 200,000 to close to two million people.[22] All of this from a war that officially didn't exist.

Then in 1971, the unsanctioned release of classified documents by Daniel Ellsberg about the Vietnam War, now known as the "Pentagon Papers," initiated a further deluge of mistrust, inundating American society. People came to know just how much they had been lied to from the beginning: lies that said the war would stop communism and lies that said victory was near. None of these were true, as the Pentagon papers clearly documented.

The ironies were many. It was, oddly enough, McNamara himself who had commissioned the secret documentation of classified war papers, resulting in the Pentagon Papers dossier, which Daniel Ellsberg, who was the Defense analyst, released. It was also McNamara, three years after his 1964 lie to Congress, who recommended in a 1967 memo to President Johnson that he consider stopping the war, declare victory, and withdraw. It was again Mc Namara, who, in another memo to the President, clearly acknowledged the suffering of the Vietnamese people, writing: "The picture of the world's greatest superpower killing or seriously injuring 1,000 non-combatants a week while trying to pound a tiny backward nation into submission *on an issue whose merits are hotly disputed* is not a pretty one" (emphasis added).[23]

Finally, in 1973, the Legislative branch, after the debacle of the Vietnam War—indirectly acknowledging their 1964 mistake—attempted to win back that critical separation of powers with the passage of the *Wars Powers Act*. It was enacted "to fulfill the intent of the framers of the United States Constitution and insure (sic) that the collective judgement of both the Congress and the President will apply to the introduction of the Armed Forces of the United States in hostilities."[24] But it only partially repaired the damage. Ensuing presidents often utilized what can only be labeled workarounds of the Act. In addition to the increased production/selling of armaments, as well as the inherent expansion of presidential privilege, McNamara's actions also loosened the existing accountability structures by stripping down and weakening the critical separation of powers doctrine in various unanticipated ways.

Rarely mentioned now, but critical to our current standing in the world, the Vietnam War was the first time the United States had overtly and formally aligned with the colonialists (this time the French) rather than with in-country liberation efforts. Our excuse was communism, but as McNamara finally admitted in his 1967 memos to the President, it wasn't a good excuse. And it could have been different.

Thirty years earlier in 1929, the Vietnamese Freedom Delegation was looking forward to meeting President Woodrow Wilson at the Paris meetings to hopefully initiate the League of Nations. Liberation groups from all of the colonies sincerely believed that the United States and President Wilson were supportive of their freedom movements—just as President Kennedy was believed to be so decades later by Algerians. As a result, representatives of the Asian and African colonies were lined up, waiting with great anticipation to meet the US President, hoping for his support.

But President Wilson would have none of it. As a staunch white supremacist, he had no interest in meeting with the so-called yellow, brown, or black people. Evidently, his liberation interests, if any, were solely focused on white East Europeans. As a result, the leader of the 1929 Vietnamese Peace delegation, who excitedly lined up to meet President Wilson in 1929, later became the famous *Ho Chi Min of North Vietnam of the 1960s*—an arch enemy of the United States. Even 30 years later in Vietnam, when initial diplomatic negotiations were simply focused on the removal of France as an occupying force of Vietnam, there could have been a feasible alternative to war. But the US Administration and the military made decisions offering military support to France that enabled war to continue—and for the first time, America stood with the colonialists.²⁵

Actually, further research indicates that this was not the first time the United States stood with the colonialists—it was just the first time we knew about it. Earlier in 1953, the Eisenhower administration, contrary to the previous Truman administration, decided to collaborate with Great Britain and remove from office a recently elected and popular Prime Minister of Iran. Doing so, they purposely strengthened the power of the Shah of Iran, until then a titular monarch with only ceremonial duties. The United States did so because the newly-elected Prime Minister was reported to be favorable to nationalizing the then British-owned oil company, thereby taking away profits that Great Britain was dependent upon. Reasons why the United States decided to become part of this coup are divided. Some say there was a fear of the communist threat. Others say it was simply nationalism vs. imperialism.²⁶ One thing is for sure. It can be argued that these two reasons together provide a hidden context for American ascendency as a world power.

The McNamara armament initiative fit right into this more intrusive international context which ironically Eisenhower had warned us against. And President Johnson, with McNamara's full cooperation, further undercut the Constitution's long-standing "peaceable bias," not just internationally but also on two domestic fronts. Because Johnson wanted to keep the funding allocated to his worthy Great Society poverty program, he did not want to ask Congress for anything that might jeopardize it. Such funding requests would notify both Congress and the American people that a real war was in the making, and necessitate increases in funding, thereby raising taxes. Equally important, it would give notice that the Reserves could be called up. So, Johnson hid war spending.

By doing so, he maneuvered around a key structure of the nation's peaceable bias—the requirement to ask Congress for war funding and the requirement to call up the Reserves, or at least put them on notice. McNamara's armament sales, not controlled by Congress but only by DOD with its proffered liquidity, were instrumental in this maneuver. But there were other possible downsides that President Johnson, consummate politician that he was, also clearly recognized. From the beginning, LBJ was hearing from congressmen and constituents making their opinions clear to him: they were "in nowise interested in having their sons' Guard and Reserve units called up to fight some godforsaken war in the jungles of Southeast Asia."²⁷ Johnson paid attention to these messages, and did a work-around. He first fought the Vietnam War with the active-duty armed forces. Only later, when forced to ask Congress for further funding, did he begin to use draftees— those young men who did not 'know someone' who could assist them in acquiring that now-coveted placement in the Reserves.

Running a war 'offline'—as the LBJ administration did—and using the active-duty soldiers until forced to call on the Reserves changed things. It altered the nature of "citizen soldier" organizations and was one of the Vietnam War's critical alterations to our country's peaceable bias. The citizen soldier organizations—the Reserves, renamed National Guards—were originally meant by the founding Constitutional framers to serve as a deterrent to war while preserving a low-key defense capability. The Reserves were also seen as organizations of patriotic citizen soldiers, ready to serve and defend their country. However, because of LBJ's unwillingness to consent to the established constitutional oversights of war established in Congress, these organizations, intended to be deterrents to war, housing patriotic citizens ready to go to war, became, instead, havens for the many that wished to avoid going to war.

As Rachel Maddow tells us in *Drift*: "for the first time, the Guard and the Reserves were the things you quietly signed up for to *avoid* service" (emphasis added). Because these Reserve enrollment slots were limited, "knowing someone" became necessary to avoid service on the front lines. Thus, the Reserves and others were corrupted in terms of their original purpose.

These were definitive mistakes, no matter how unintended the consequences. The arms sale precedent and our actions in Vietnam supporting France's colonization weakened our nation's peaceable bias and weighted our international foreign policy toward greater use of force and power. It set us on a different, and perhaps unintended path.[28] But whatever the intent, the era of the expanding military complex without sufficient citizen oversight had begun.

But still, these three historical events—becoming the arms merchant of the world with an aversion to audits; the expansion of presidential power to wage war; and the corruption of the citizen soldier concept—all with the subsequent erosion of the Constitution's peaceable bias—were, by themselves, insufficient to create an actual *practice* of *endless war*. They were still operating within the established norms but pushing at the edges. There are, however, two other events, one overt and one covert, which are critical to the actual practice of endless war.

Playing Outlaw

It was perfect timing for the advent of a new incoming Reagan Administration, with its affinity for executive power expansion. Hints of this power expansion were first noted some 15 years earlier, during the Vietnam War, by Senator William Fulbright, Chairman of the Senate Foreign Relations Committee. In his book, *The Arrogance of Power*, Fulbright presciently captures the key elements put in play during the 1960s, and then massively expanded during the incoming Reagan administration. Fulbright observes that "in the course of dehumanizing an enemy—and this is the ultimate fallout of any war—a man dehumanizes himself. It is not just the naturally bellicose ... it is everyman." Fulbright continues:

> America is showing some signs of that fatal presumption, that overextension of power and mission, which has brought ruin to great nations in the past. The process has hardly begun, but the war which we are fighting now can only accelerate it .... If that fatal process continues to accelerate until America becomes what she is not now and never has been, a seeker after unlimited power and empire ... then Vietnam will have had mighty and tragic fallout.[29]

Moving forward 20 years, to understand Reagan's impact on our country, we must look at his authoritarian actions and not his affable words; these words were many and sometimes even inspiring (think shining city on a hill). But the Republican conservatives serving in his administration were not interested in shining cities, particularly those serving on the National Security Council who undertook those subsequent well-planned corrupt acts. Instead, these executives recognized with great clarity the advantages of increased presidential executive power. They understood the resulting free rein concerning armaments and budgets within the military and the advantages of the international global policeman role being offered to them. All of this became possible because of the increasingly unchecked expansion of executive power and the resulting privatization of Reagan's foreign policy initiatives. Together, the pairing of power expansion and privatization of initiatives allowed the White House to illegally run its Nicaragua operation, clear of congressional constraints and any kind of organized accountability.[30]

Reagan's preference for executive power, with its lessening accountability to Congress and the American people, was also quite effectively juxtaposed next to his favorite mantra, "*Government is the problem.*" This mantra fit perfectly with the decreasing levels of trust in government experienced by the American people since the loss of the Vietnam War and the nation's then-current economic problems. Together, they were indicators of increasing toleration for a non-accountable government, which in turn allowed the increased expansion of executive, if not authoritarian, power for war, as predicted by Senator Fulbright, and first initiated in the Iran-Contra Affair. But because these actions took place at the highest levels of government, the true criminality of the Contra affair took more than a decade to prove. It was finally revealed in 1988, despite intervening presidential pardons. But this implementation of enhanced executive power also began the diminution of democracy, which continues today. That is why this history remains so relevant.

Reagan's administration, first working legally through the CIA, surreptitiously used Honduras as a staging area, handing over more than $20 million to bribe Honduran generals and train troops. Honduras was just one of several Central and South American countries willing to help Reagan as long as their own extra-legal coffers were replenished. These countries included Honduras, Guatemala, El Salvador, and the Contras of Nicaragua—all countries, it should be noted, strongly contributing to today's massive immigration to the United States.

It is well documented that the Reagan administration, and the President in particular, held the Nicaraguan drug-dealing fighters in high esteem. The President himself called them "freedom fighters" because they were battling the pro-left but legally elected government of Nicaragua. Initially, through calculated moves using enhanced "executive freedom," the Reagan Administration legally began to offer militarily training support to the right-wing, drug-dealing Nicaraguan guerilla forces known as the "Contra's."[31]

The Iran-Contra affair didn't begin as a criminal activity. But when it was revealed that the CIA was physically aiding the Contras in actual sabotage, Congress once again attempted to regain its primary war powers by passing the Boland Amendment in 1982-83. This time the amendment to curtail war specifically prohibited the federal government, meaning the Reagan administration, from providing military support "for the purpose of overthrowing the Government of Nicaragua," taking away the supposed legality of the Administration's Contra initiatives.

But the Reagan Administration decided to continue *illegally* assisting those Contra anti-government forces. No attention whatsoever was paid to the fact that the US Congress had expressly and specifically outlawed any military assistance to Central America, fearing another Vietnam-type involvement. Directing these illegal moves from the White House Executive Branch in the 1980s, Colonel Ollie North and a small, strategically placed group within the Executive offices of the White House repeatedly undertook three illegal actions: (i) they lied to Congress; (ii) they circumvented the in-place congressional Boland Amendment ban on financing the Contra war in Nicaragua and (iii) they worked undercover with known drug cartels to accomplish their goals.[32]

Evidently, from the beginning of his first regime according to numerous reports, Reagan himself was convinced that he had unassailable authority over all aspects of national security, "claiming the right to go to war, in secret, against the express will of Congress."[33] Thus, the Reagan administration, once legal funding was no longer available for funding the Contra's from the US government and private donors, decided to utilize funding from their ongoing *illegal armament sales to Iran.*

This illegal arms sale to Iran was initiated by the Reagan Administration in 1981, despite a clear arms embargo against Iran. It was only stopped in 1986 when these illegal shipments, including more than 200 TOW Anti-tank missiles, were uncovered as part of the House Select Committee investigating the Iran-Contra Affair. These arms sales were going to the regime, headed by Ayatollah Khomeini, President of Iran which had approved the Iran-hostage debacle where 53 American diplomats were held against their will for 444 days from 1979 to 1981. With these revelations in 1986, Americans were, for the most part, horrified, and Reagan's ratings fell more than 30%. This same extreme-right regime remains in power today under the rule of a second Ayatollah.

In 1988, the Congressional Committee's Report on the Iran-Contra Affair explained reaction to these revelations.

> To many Americans, the very idea of selling advanced weaponry to the Ayatollah Khomeini was, all by itself, more than enough to send the President's approval ratings into a steeper, faster free-fall than the polls had ever registered since they were first taken when Franklin Roosevelt sat in the White House .... How much did the President know about all of that? ... In October of 1986 he said he "did not know the exact particulars" of the Contra supply program. But then in May he said the whole thing "was my idea." That contradiction has never been explained.[34]

However, while these Iran-Contra news stories were clearly out in public during the 1980s, Reagan's Hollywood-enhanced political popularity held charges and justice efforts at bay. This status was enhanced by Reagan's Federal Communication Commission's abolishment of the *Fairness Doctrine*, which had previously mandated that broadcasters provide balanced coverage of controversial issues of public interest. Without this mandate in place, the one-sided right-wing talk radio with Reagan's focus on "government as the problem" took off with no counterbalance, and continues today.[35]

As a result, it was only at the insistence of Senator John Kerry, then a newcomer to Congress, that the "Report of the Congressional Committees Investigating the Iran-Contra Affair" was completed.[36] But even so, the riveting 1989 Congressional investigation report about the evidence of CIA-backed Contras was downplayed and neglected by the national news media. One investigator reported:

> Although Kerry's findings represented the first time a congressional report explicitly accused federal agencies of willful collaboration with drug traffickers, the major news organizations chose to bury the startling findings. The New York Times, the Washington Post, and the Los Angeles Times all wrote brief accounts and stuck them deep inside their papers .... Thus Kerry' reward for his strenuous and successful efforts to get the bottom of a difficult case of high-level corruption was to be largely ignored by the mainstream press and even have his reputation besmirched.[37]

Here is one more often forgotten item about the Contra scandal that is pertinent to our current history-making. Independent Counsel Lawrence

Walsh, who was appointed to the federal bench by President Eisenhower and then directed the CIA Independent Counsel investigation in 1986–87, courageously did bring indictments against the high-level implicated Reagan appointees for those Contra activities. As a result of the investigations, Walsh indicted the high-level Reagan and Bush appointees who were involved: Secretary of Defense Caspar Weinberger, national security advisors Robert C. McFarlane and John Poindexter, and Assistant Secretary Elliot Abrams.

Although they were all charged—as directed by Independent Counsel Walsh—President H.W. Bush, who had succeeded President Reagan and who was also at least partially implicated, decided otherwise. President Bush pardoned them all on Christmas Eve, 1992. Mr. Walsh summarized the effect of these actions: "pardons by President H.W. Bush effectively ended the prosecutions and affected a final layer of coverup over the whole affair."[38] So, did it end there? No, not exactly. In 1998, almost a decade later, Kerry's hard-to-believe investigation was finally vindicated, somewhat unwillingly, by no-less than the Inspector General of the CIA, Frederick P. Hitz. This official US Government report found *"that scores of Contra operatives were implicated in the cocaine trade and that US agencies had looked the other way rather than reveal information that could have embarrassed the Reagan–Bush administration."*[39] But there is one more question. It's interesting to ask ourselves a telling question: who was the assistant US Attorney General who so willingly carried out these pardons of the Iran/Contra conspirators for then President Bush? Do you remember? Yes, now we all remember. The same man who was appointed Attorney General by President Trump in 2019—William H. Barr.

Now the negatives of this accountability loss keep rolling in, and we seem to have no real answers. The collaboration of the Congressional MAGA Republicans with Putin's Russia, refusing to support the Ukrainian War, illustrates how far this loosening of our nation's accountability has taken us in this 2020 decade. We essentially allowed an unchecked expansion of the military-industrial complex over the decades, as Eisenhower warned against. But evidently when needed, those King's men currently operating under the MAGA flag do their best to block this munitions production to be used against the authoritarian enemies of democracy.

So horrific and unjust wars continue. In 2023–2024, the horrific Hamas invasion of Israel with its violent and brutal killing and taking of hostages, was followed by the equally horrific bombing retaliation against Gazan civilians by Israel's Netanyahu government. The chief of the United Nations has recently announced this bombing of civilians to be in "clear violation" of humanitarian law. Nor can we forget—the wars in such places as Sudan, rarely talked about, where in 2024 mass starvation was imminent because two armies were supplied with guns, planes, and bombs to fuel their fight over oil. The stupidity and immorality of all of this basically prostrates us all. However, when we look at the history squarely, however, I believe there is one basic fact that can be agreed upon. That basic fact—accountability is now cast aside—means that *endless war* and *complicity reign!* And that complicity encompasses everyone. But because of this long-term, unacknowledged, ongoing complicity, I also see that "hope in the dark" is the one facet of this encompassing disaster, which currently includes Israel, Palestine, Ukraine, and Sudan which might save us all.

"Hope in the dark" is the relational hope of recognizing connection: the hope of ripping apart the previous racial, ethnic, nationalistic, and power norms by those numerous groups who are saying ENOUGH! It's the Jewish and Arab university students demonstrating together to say "no" to continued American support of Israel's Netanyahu bombing. It's the Russians who commemorate Navalny's name with their brave public demonstrations against Putin. It's all of us who understand that because democracy requires not only freedom but also a basic equality of community, it necessitates that we leave aside the old ideas of divisive living, and instead build new ways of interdependent belonging.

Looking back, we begin to understand. This nefarious remake of our Constitutional separation of powers into a unitary executive theory of power—with its loosening of accountability to the people—begun by McNamara and LBJ; institutionalized by Reagan and his cohort; continued by H.R. Bush, George W. Bush and Cheney, and then Trump, has had serious results. It has continued to expand non-accountable power, tolerate corruption, decrease citizen trust in democracy, deny justice, and facilitate war. As a result, this corrupting ideology—placing the executive power of action with the legislative power of the "purse" together with no controls whatsoever—has clearly weakened our democracy, enhanced for us the potential implementation of an authoritarian government, and expanded the possibilities for endless war around the globe.

The War at Home

But what has been happening at home, in our own country, during all of this time? In retrospect, all of us have been left with hardly a clue for the past fifty years or so. All we had was a fuzzy background landscape and no telescope. As a result, a plethora of violent, separate domestic incidents, were able to pass by without connection or definition of pattern: militias, Ruby Ridge, Waco, Oklahoma City, Columbine, Sandy Hook, Parkland, Uvalde, and so many more. It was disturbed young men taking guns to school and executing their schoolmates; one deciding to murder 15 people "to start a race war"; another deciding to commit mass murder in a synagogue; and intermittent but continuing police murders of unarmed black women, men, and boys. All of these were described, over and over, as unfortunate but separate incidents. But now we begin to see the connections and the patterns of violence and war at home, in our cities, and in our small towns and rural areas.

Despite these extremely difficult possibilities, remember, I stated earlier, that there were still *two further events* needed to completely establish a national practice of endless war. One was the just-discussed overt outlaw actions of the Reagan and subsequent Bush administrations, which further loosened the peaceable bias of accountability and allowed corruption a way in. The second idea and its subsequent series of events, still then covert, would expand "government is the problem" into the necessity for revolution. That second idea, no longer covert, has now exploded into the war at home. We tend to blame it on Trump, but he is basically just taking advantage of this larger phenomenon that Katherine Belew and others outline for us.

Belew begins her powerful book, *Bring the War Home: The White Power Movement and Paramilitary America*, with a new reality that we are only now beginning to recognize. This little understood pattern of war was instigated by Americans ... against Americans. The underlying philosophy and belief system ensuring a practice of endless war, now being widely disseminated here at home has a history. And if, and when, this history is finally successful, dissemination of this underlying philosophy is expected to result in white authoritarian rule. But we're not there yet. However, this low-key pattern of war relying on divisive racial, ethnic, nationalistic, and power norms is having success in creating deep cleavages among us, which is a primary step toward civil war.

Reading Belew's book, with its impeccable research, is like stepping up to a telescope and focusing on a fuzzy background landscape—immediately long-forgotten and misunderstood elements become discernible, even well defined. She describes a white power and paramilitary movement, emphasizing its multiple connections, interactions, and violent successes. Until recently, much of this has been missed by all of us, including the FBI. But Belew's research effectively documents the planned recurrence and expansion of chaos-inducing violence that presages acceptance of authoritarianism, now most recently laid out step-by-step in "Project 25," the proposed governing document for MAGA Republicans.

Belew begins her book with the following:

"WE NEED EVERYONE OF YOU," proclaimed an anonymous 1985 article in a major white-power newspaper. "We need every branch of fighting, militant whites. We are too few right now to excommunicate each other .... Whatever will save our race we will do!" The article spoke of emergency and government treachery. It foretold imminent apocalyptic race war. It called to believers in white supremacist congregations, to Klansmen and southern separatists, and to new-Nazis. The white power movement united a wide array of groups and activists previously at odds, thrown together by tectonic shifts in the cultural and political landscape. Narratives of betrayal and crisis cemented their alliances ....

This malevolent invitation was initially instigated, oddly enough, by a small group of returning Vietnam military veterans, turned white supremacist advocates. They saw *difference* not only as a negative but as a mark of unworthiness to be expunged with brutal violence, focused first on blacks, and secondly, anyone associated with communism. Kathleen Belew tells us that at its inception in 1979, this small group of returning Vietnam veterans rallied to the cry to "bring the war home." This meant expunging blacks who were considered to be either "dupes," or representatives of communism. Violence, particularly "race" violence, as these white supremacists initially saw it, was the primary organizing principle in 1979.

But the white-power movement also had larger goals. The real work for those militants became, and remains so now, the building of a unified and violent white power movement at home and abroad. As a result, in 1983, only four years later, the white-power movement with this small group of returning Vietnam veterans at its center did an abrupt about-face: they declared war against the federal government; even in the midst of the Reagan presidency, which they had previously supported. Belew does document their strong support of Reagan's Contra policy, even to the point of some volunteering to fight in South America with the Contras. So, it is interesting to note that this abrupt about-face of this movement came in the same year that the Boland Amendment began to be enforced.

While reactionary politics and a wished-for return to a Jim Crow society characterized the old-style Klan-type organizations, the new white power movement now sought something very different: "revolution and separation—the founding of a racial utopian nation."[40] There were two strategies—civilian disruption and police/security infiltration—to be carried out under the strategy of *leaderless resistance*. This establishment of "leaderless resistance"—directing small individual cells of one or more people—to act without contact or specific direction from the movement leadership turned out to be nothing less than brilliant. It was designed to define each violent bombing or killing as an isolated event. It also made it extremely difficult to trace any line of responsibility beyond the lone perpetrator. And so, it continues today. In retrospect, "leaderless resistance" implemented against both the Federal Government and multiple communities, was and is, critically important; but it remains a rarely recognized tool of the White Power movement even today.[41]

A major accompanying objective which has strengthened the war at home is *paramilitary policing*. The militarization of national, state, and community police forces is rarely mentioned by the media nation-wide. But because it has been given strong emphasis in the white-power movement, it accomplished paramilitary policing goals on three fronts. First, the declaration of war on the national state in 1983 by White Power groups served to initiate the now almost forgotten militant standoffs with the federal and state governments (think Ruby Ridge and Waco). And it was these standoffs that provoked traditional police enforcement to move toward paramilitary enforcement. Second, in the 1990s, the US military took advantage of a rise in violence to convince local police forces to 'take' and 'use' the military's oversupply of weapons and equipment. This effectively created the mini-militarized police SWAT teams that we now see across the country. Third, numerous white power members joined police forces in every state, introducing their White Power beliefs and expanding more violent methods of police action, particularly against urban black and brown communities. Belew provides statistics, and I have to admit, these military weaponization statistics that she outlines below, makes me think back to the story of Tyre Nichols in Chapter 2 and his meeting with the local SWAT team.

> After the war, military training of police departments and paramilitary units such as Special Weapons and Tactics (SWAT) teams brought violence home. Paramilitary police units … would grow exponentially, even as federal police agencies, including the FBI, ATF, and DEA, militarized along the same lines. Almost 90 percent of cities with 50,000 or more residents would have paramilitary police units by 1995. The use of such units would grow 538 percent for "call-outs" (responses to emergency service calls) and 292 percent for "proactive patrols" (including the suppression of communities of color) between 1982 and 1995.[42]

The bombing of the Oklahoma City Federal Building on April 19, 1995, by Timothy McVeigh exactly two years after the Waco siege, killing 168 and wounding more than 500, also played an important, but now almost forgotten, role in bringing the war home. The "leaderless resistance" strategy initiated by the white power movement should have alerted the US security and policing apparatus and the American people as well. Instead, theories promulgated after the Oklahoma bombing focused on the *lone-shooter theory* who had mental health issues. As a result, Belew explains:

> [T]he disappearance of the movement in the years after Oklahoma City—engineered by white power activists but permitted and furthered by government actors, prosecutorial strategies, scholars, and journalists alike—left open the possibility of new waves of action.[43]

And so it did. The 2015 shooting by Dylan Roof of nine black bible study attendees at the Emanuel African Methodist Church in Charleston, South Carolina—because he wanted to "start a race war"—was one of the many increasing lone-gun shootings. Like McVeigh, he clearly had no mental health issues other than hate. However, despite similarities in terms of intent and procedure, there was one large difference between McVeigh and Dylan Roof.

While McVeigh was a militia member, attending gun shows in order to associate with like-minded white-power people, Roof only had to go to his preferred alt-right website to learn how to follow the white-power movement's teachings.

So even now, the so-called lone and leaderless shootings continue to accumulate with all of their terror, misery, and sorrow. Rachel Maddow interviewed Kathllen Belew on her February 6, 2023, TV program, where they discussed, in some depth, continuing domestic terrorism built on the belief in white supremacy. Perhaps what sums it up best is that death by guns is now the number one reason for the death of American children under the age of 13. Given the fact that even the smallest bit of pushback requires enormous effort, indicates that this belief system which allows our children, loved ones, friends, and community members to be sacrificed in the name of guns and violence may be taking hold.

Two other books give us further information. *Everything You Love Will Burn* by Norwegian writer Vegas Tenold documents the rebirth of brutal violence in white nationalism. The most recent book, Amy Cooter's *Nationalism, Nostalgia, and the US Militia Movement*, explores the inner components of what makes US militias a real danger to US democracy and not just "social outcasts on the fringes of society."

It should be noted, however, that former President Trump fully recognized this undercover movement from the start and continues to use it to his great advantage. Both his previous administration and his 2024 campaign to be president were not "off the wall" as many have believed. Instead, Trump continued to aim exactly at his chosen audience—and it definitely includes and focuses upon his white-power constituents. And, perhaps most importantly, we must begin to understand that Trump's love of Putin is directly aligned with the international White Power movement, for which Putin is the nominal leader.[44]

But finally, this little-recognized militarized expansion, with its accompanying domestic terrorism focused on activating fear and hate, is slowly being recognized and explored by our Federal Government. But certainly not adequately—they are still tiptoeing around it. So far, the Southern Poverty Law Center and the Anti-Defamation League are the only two organizations which have truly kept the White Power movement and other hate organizations in view and in check through lawsuits and continued reporting. Will our government take on a more substantive role? That remains to be seen.

Sanctification: Guns, War, and Violence

It is here that we must understand a difficult reality. I must admit that when I began to perceive this pattern, I experienced a sense of sick sadness—a queasiness—that was new to me. As Katherine Belew points out in the final paragraphs of her book, "the lack of public understanding, effective prosecution, and state action for terrorist acts fueled by belief in white supremacy left the door open for continued transgressions." Of course, the question underlying this state of affairs is "why," and Belew purposely leaves that question to all of us as readers. My first admittedly optimistic thought was: maybe we just don't know all the facts; maybe there are real undercover ongoing investigations. At least that would be a start.

But then I found a report, after some searching, that did have all the facts, or at least many of them. In November 2022, the US Senate Committee on Homeland Security and Governmental Affairs issued its report on domestic terrorism after three years of investigation. The report focuses on "the rise in domestic terrorism, the federal response, the allocation of federal resources to addressing domestic terrorism, and the role of social media companies in the proliferation of extremist content."

The Committee sent document requests, held multiple interviews, and reviewed over 2000 key documents. The most critical findings were the following:[45,46]

- "Since the federal government shifted its focus predominantly toward international terrorism [in 2001], attacks from domestic terrorists have surged. According to a 2021 Center for Strategic and International Studies study, there were 110 domestic terrorist plots and attacks in 2020 alone, a 244 percent increase from 2019."
- "Although outside researchers have reported on trends relating to domestic terrorism, the federal government has not systematically tracked and reported this data itself, despite being required to do so by law."
- "This investigation found that the federal government has continued to allocate resources disproportionately aligned to international terrorist threats over domestic terrorist threats."

It was after reading this US Senate report and asking myself *why* that I first suffered that first onslaught of queasiness. Yes, we now have the January 6 Committee hearings and the 2023 DOJ indictments of former President Donald Trump. But this Senate Report indicates a profound and continued reluctance by people in the federal government to investigate domestic terrorism—even as the organizations themselves recognize this threat may supercede the international threat. So, what's going on here?

As I searched for plausible explanations for this troubling pattern outlined by the Senate Report, I remembered Derrick Bell, the noted civil rights lawyer and professor. His analysis of racial discrimination and why legal efforts to combat it so often failed, as briefly mentioned in Chapter 2, seemed to offer a possible mode of inquiry. In his famous book, *Faces at the Bottom of the Well*, first published in 1992, Derrick Bell points out that democratic civil rights law, based on his long experience, often does not work as intended.

He explains that "based on our agreed-to law enforcement model, it assumes that most people will obey and socially enforce the laws." Bell points out that "this model breaks down when a great number of whites are willing—because of convenience, habit, distaste, fear, or simple preference to violate the law." He goes on to observe, "It then becomes almost impossible to enforce, because so many whites, though not discriminating themselves, identify more easily with those who do, than with the victims."[47]

If we are, however, to effectively consider Bell's argument in situations other than race, a basic understanding needs emphasis. Bell alludes to it, but does not spell it out. That is, "democracy depends upon *social enforcement*." We don't often think about it, but authoritarian nations require oppressive policing to ensure that their laws are enforced—people will break them when given a chance. However, democratic nations do not require this oppressive force because the people agree with the laws and abide by them with respect. This means that the majority of laws are enforced on a day-by-day basis by social consensus and social enforcement.[48]

I bring Bell's perspective, that the law enforcement model doesn't work, forward to suggest that a similar sickening phenomenon may be underwriting our difficulties in coming to terms with both endless war and domestic terrorism. Do numerous employees and leaders in our national security agencies have strong difficulties in truly comprehending that people who look like them, who claim to be ardent Christians, perhaps like them, are actually domestic terrorists who commit atrocious acts of violence? Is that possible? Is this also the reason that security agencies have continued their concentration on international terrorists, who are often brown and Muslim, rather than domestic terrorists, who are often white and Christian? Are they willing, "because of convenience, habit, distaste, fear, or simple preference," to *ignore* enforcement of the law?

Or, is something else going on? Is it also possible that our national security agencies have not yet sufficiently investigated these phenomena because they understand that a sizeable proportion of the population is cognizant of this white power activity and *tolerates* it? So, are the security agencies just going along to get along? Is there a lack of *social enforcement* for such an investigation on both sides "because of convenience, habit, distaste, fear, or simple preference?" Should we consider this as a second option? Sadly, I believe both of these perspectives are possible.

So, my queasiness persists. We now understand that ever since the 1960s, we have moved away from our country's intentional "peaceable bias"—step by step—thereby providing the backdrop for authoritarian beliefs and violent action. The presidential capacity of the Executive Branch to act in secret and without accountability was expanded (think of the State Department under Reagan, the DOD under George W., and the Department of Justice under Trump). And now, there is also this endless war syndrome here at home—up-close mass murders supposedly perpetrated by loners, supported by the white supremacy network's promotion of addiction to chaos and fear. This situation is backed by the armament profit makers, who have sold more than 400 million guns in the United States, normalizing violence and sanctifying fear, with the help of some Congressional representatives.

These are all the components that accomplish authoritarian goals. So, all talk and façade of democracy aside, this is the culture within which we are now operating—however uneasily. Full acceptance of this new organizing principle of violence and its resulting culture will overtake everything—unless we think it through and act fast. Why? Because if belief in the inevitability of violence is switched *on*—we no longer see the alternatives—although these alternatives do exist—violence rapidly becomes the central organizing factor of society. This reorganization then leads us down the uncharted path of *endless* war, a pattern that has been verified over and over. Can we change that process? That remains to be seen. But I hope so; I believe so.

Notes

1. John Ralston Saul. *Voltaire's Bastards: The Dictatorship of Reason in the West* (New York: Vintage Books, 1992) 141–176.
2. Edward J. Drea, *McNamara, Clifford, and the Burdens of Vietnam: 1965-1969* (Washington D.C: Secretaries of Defense, Historical Series, 2011), Vol. VI.
3. Daniel Immerwahr. *How to Hide an Empire: A History of the Greater United States* (New York: Farrar, Straus, & Giroux, 2019. For those interested in further exploration of this hidden history, this book is worth reading in its entirety.
4. Immerwahr, *How to Hide an Empire*, Chapter 18.
5. Immerwahr, *How to Hide an Empire*, 391, 392.
6. Marine Corps. *Small Wars Manual* (United States Marine Corps, 1940–2018). Sunflower University Press, Manhattan, Kansas, 1992.
7. Colman McCarthy. *I'd Rather Teach Peace* (New York: Orbis Books, 2002), 20.
8. Immerwahr, *How to Hide an Empire*, 17.
9. Danile Brown, "SIPRI data base: Trends in International Arms Transfer", *Business Insider*, Mar. 16, 2018, https://businessinsider.com.
10. David Lindorf, "Special Report: Exposing the Pentagon's Massive Accounting Fraud", *The Nation*, November 2018, 13. https://thenation.com.
11. National Archives. "President Dwight D. Eisenhower's Farewell Address (1961). https://archives.gov.
12. Lindorf, "The Pentagon's Massive Accounting Fraud", 13, https://thenation.com.
13. Lindorf, "The Pentagon's Massive Accounting Fraud" 14, https://thenation.com.
14. Lindorf, "The Pentagon's Massive Accounting Fraud" 15, https://thenation.com.
15. 'The balance' offers a simplified look at the US economy, 2019, https://thebalancemoney.com.
16. Immerwahr, *How to Hide an Empire*, 400.
17. Immerwahr, *How to Hide an Empire*, 390.
18. Alexander Hamilton, James Madison, John Jay. *The Federalist Papers* #51 (London: Arcturus, 2016).

19 Rachel Maddow. *Drift: The Unmooring of American Military Power* (New York: Broadway Paperbacks, 2012), 8.
20 Elizabeth Becker. "The Vietnam War, Exposed in One Epic Document", *New York Times*, Aug. 2021, https://nytimes.com. This article is excellent, covering the historical period and impact of the Pentagon Papers.
21 Becker, "The Vietnam War", https://nytimes.com.
22 There are numerous online sites that discuss the number of deaths in detail, ranging from Brittanica to Wikipedia. https://britannica.com, https://wikipedia.org.
23 Becker, "The Vietnam War", https://nytimes.com.
24 War Powers Act. S.440, 93rd Congress, 1973-1974, https://congress.gov.
25 Pankaj Mishra. *Bland Fanatics: Liberals, Race, and Empire* (New York: Farrar, Straus, and Giroux, 2020), 74.
Daniel Immerwahr. "You Can Only See Liberalism from the Bottom Up", *Foreign Policy*, Sept. 2020, https://foreignpolicy.com.
26 Ervand Abrahamian. "The 1953 Coup in Iran", *Science and Society*, 65 (2001) Vol 65, No.2, 2001.
27 Maddow, *Drift*, 16–17.
28 Maddow, *Drift*, 16.
29 Fulbright. William, *The Arrogance of Power* (New York: Random House, 1966), 138.
30 Inouye, Daniel K. and Lee H. Hamilton. *Iran Contra Affair: Report of the Congressional Committees Investigating* (New York: Random House), 298, 303, 352, 361.
31 Inouye and Lee, *Congressionnal Report : Iran Contra Affair*, 1988, Iran-Contra Chronology, xv.
32 Inouye, *Congressionnal Report : Iran Contra Affair*, 1988, Iran-Contra Chronology, 11–34.
33 Maddow, *Drift*, 113.
34 Inouye and Lee, *Congressionnal Report : Iran Contra Affair*, 1988, Introduction, x-xi.
35 Ronald Reagan Presidential Library, "Fairness Doctrine", https://reaganlibary.gov.
36 Daniel K. Inouye, Senate Select Committee Chair and Lee H. Hamilton, House Select Committee, "Report of the Congressional Committee Investigating the Iran-Contra Affair", NY 1988. (Currently available on Amazon.)
37 Robert Parry, "How John Kerry Exposed the Contra Cocaine Scandal", *Salon*, Oct. 25, 2004, 10. https://salon.com.
38 Lawrence Walsh, *Firewall: The Iran-Contra Conspiracy and Cover-up* (New York: W.W. Norton, 1998).
39 Parry, "How John Kerry Exposed the Contra Scandal", *Salon*, Oct. 25, 2004, 2. https://salon.com.
40 Belew, *Bring the War Home*, 5.
41 Belew, *Bring the War Home*, 4-5.
42 Belew, *Bring the War Home*, 189.
43 Kathleen Belew, *Bring the War Home*, 237.
44 Numerous scholarly articles have been published on international white nationalism in the past decade, particularly in the New York Review of Books.
45 This report also includes information on social media not included here.
46 US Senate Committee on Homeland Security and Governmental Affairs, "Rising Threat of Domestic Terrorism: Further Action Needed", Nov. 2022, executive summary, https://hsgac.gov.
47 Derrick Bell, *Faces at the Bottom of the Well: The Permanence of Racism* (New York: Basic Books, 1992, 2018), 55–56.
48 I developed and wrote about this concept of social enforcement while working with local and regional groups organizing participatory and research initiatives in several conflict-affected countries.

Bibliography

Cooter, Amy. *Nationalism, Nostalgia, and the US Militia Movement.* New York: Routledge, 2024.
Bell, Derrick. *Faces at the Bottom of the Well: The Permanence of Racism.* New York: Basic Books, 1992.
Irvin Painter, Nell. *The History of White People.* New York: W.W. Norton, 2010.
Isenberg, Nancy. *White Trash: The 400-Year Untold History of Class in America.* New York: Penguin Books, 2017.
Piketty, Thomas. *Capital and Ideology.* Cambridge: Belknap Press of Harvard University, 2020.
Tenold, Vegas. *Everything You Love Will Burn: Inside the Rebirth of White Nationalism In America.* New York: Nation Books, 2018.
Waite, Kevin. *West of Slavery: The Southern Dream of Transcontinental Empire.* Chapel Hill: Univ of North Carolina Press, 2021.
Zinn, Howard. *A People's History of the United States.* New York: Harper Modern Classics, 2015.

6
SUFFERING OF POVERTY

The status of poverty in the United States cannot be sugar-coated. Its reality runs contrary to everything we have believed about ourselves as a country. To review, our overall situation is best illustrated by comparing our pre-2020 Covid-19 status to other countries of similar economic status. A 2017 assessment and report by the "United Nations Special Reporter on Extreme Poverty and Human Rights," compiled by the Organization for Economic Cooperation and Development (OECD), of which the United States is an organizing member, gives a reality-based soul-shaking assessment.

Rich-Country Poverty: Facts and Status

To begin, the United Nations OECD Report provides a series of up-to-date comparisons of US poverty status in contrast to other similarly economically placed countries.[1] Here are the basic facts.

- By most indicators, the United States is one of the world's wealthiest countries.
- In the OECD, the United States ranks 35th out of 37 in terms of poverty and inequality.
- US inequality levels are far higher than in most European countries.
- According to the World Income Inequality Database, the United States has the highest Gini rate (measuring inequality) of all Western countries.
- US healthcare expenditures per capita are double the OECD average and much higher than in all other countries. But there are far fewer doctors and hospital beds per person than the OECD average.[2]

Philip Alston, the UN Special Rapporteur, author of this report, and the United Nations watchdog of extreme poverty worldwide, issued what Reverend Dr. William J. Barber II, leader of the Poor People's Campaign in the United States, described as "a withering critique of the state of America today." The Reverend goes on to observe: "policies that benefit the rich while deregulating companies and neglecting the poor are steering the country toward a dramatic change of direction, blocking poor people from accessing even the most meager necessities." Alston himself concludes that while he saw much that was positive in his visiting assessment, "instead of realizing its founders admirable commitments, today's United States has proved itself to be exceptional in far more problematic ways that are shockingly at odds with its immense wealth and its founding commitment to human rights. As a result, contrasts between private wealth and public squalor abound."[3]

Sometimes an outsider perspective is most revealing. Alston's answer to, and description of, how we in America describe "who are the poor?" is most useful, and begins to describe exactly where we are.

> I have been struck by the extent to which caricatured narratives about the purported innate differences between rich and poor have been sold to the electorate by some politicians and media, and have been allowed to define the debate. The rich are industrious, entrepreneurial, patriotic, and the drivers of economic success. The poor are wasters, losers, and scammers. As a result, money spent on welfare is money down the drain. To complete the picture we are also told that the poor who want to make it in America can easily do so: they really can achieve the American dream if only they work hard enough.

> The reality that I have seen, however, is very different. It is a fact that many of the wealthiest citizens do not pay taxes at the rate that others do, hoard much of their wealth off-shore, and often make their profits purely from speculation rather than contributing to the overall wealth of the American community. Who then are the poor? Racist stereotypes are usually not far beneath the surface. The poor are overwhelmingly assumed to be people of color, whether African Americans or Hispanic 'immigrants.' The reality is that there are 8 million more poor Whites than there are Blacks in poverty. Similarly, large numbers of welfare recipients are assumed to be living high on the hog ... But the poor people I met from the 40 million living in poverty were overwhelmingly either persons who had been born into poverty, or those who had been thrust there by circumstances largely beyond their control such as physical or mental disabilities, divorce, family breakdown, illness, old age, unlivable wages, or discrimination in the job market.[4]

Alston makes two additional observations that are of particular importance. The first is that in an extremely rich country like the United States, *"the persistence of extreme poverty is a political choice made by those in power."* So let's read that sentence again ... it's not the poor who sadly suffered bad luck, or even more punitively, just didn't work hard enough? Instead, it is a "choice" made by those in power? Not too many people say it so bluntly, do they? But sometimes "blunt" is good.

Consequences

Poverty has always been endemic in the United States; as the previous chapters on the injustice of racism, sexism, and classism clearly illustrate. But Philip Alston, in the same report, also outlines current national consequences based on our factual situation, as compared to other countries.

- The youth poverty rate in the United States is the highest across the OECD with one-quarter of youth living in poverty compared to less than 14% across the OECD.
- The Stanford Center on Poverty and Inequality characterizes the United States as "a clear and constant outlier in the child poverty league." US child poverty rates are the highest amongst the six richest countries—Canada, the United Kingdom, Ireland, Sweden, and Norway.
- US infant mortality rates in 2013 were the highest in the developed world.
- The United States has the highest prevalence of obesity in the developed world.
- Americans can expect to live shorter and sicker lives, compared to people living in any other rich democracy, and the "health gap" between the United States and its peer countries continues to grow.[5]

The structure of poverty in the USA, according to numerous recent data studies has also gone through dynamic changes over the past two decades. However, these changes have received little attention except from poverty researchers. The authors of "The Facts Behind the Vision" indicate that the recognition and acceptance of high national levels of *inequality* is part of the reason for the little attention paid to these changes. But they also admit to the worry that *if* there is an open and frank discussion about the reasons for these changes it might be "counterproductive because some reformers (read politicians) might seize on that discussion to justify reforms oriented more to reducing spending than reducing poverty."

Obviously, this statement intensifies the sense, mentioned above, that something is missing in our current understanding of poverty; and frankly, when I first read their statement it gave a jolt to my heart—are the researchers telling us they are afraid to say the truth because it could be used in a vengeful way? But no matter, we are indebted to these poverty professionals who tell it exactly like it is. This is what they say.[6]

> We offer this article in the admittedly quaint hope that it is better to operate with full and complete transparency and that an open and honest discussion of the facts will in the end lead to informed poverty-reducing policy. The simple predicate of this piece that given the massive externalities brought on by running a high-poverty economy, there is an open and shut case for reform efforts that are authentically focused on *reducing* the poverty rate. We will attempt, therefore, to identify the key poverty facts that such legitimate reform efforts should bear in mind. (emphasis is theirs)[7]

Before we move forward to the facts and evidence, I am going to be honest and say that I had to read this Pathways publication twice, and then rewrite this chapter. Basically, the first time, I was reading for information; the second time I finally got it—it's also the *meaning* behind the words—the overall meaning. The paragraph above tells it all. In effect, this article is the scientific, restrained, professional act of ethical human beings marooned on an island of expertise that is too often not taken into account. They are surrounded by a clear danger that they know is hurting myriads of people; so, they are throwing a bottle in the ocean with a message that says to the rest of us: DO SOMETHING—we've done everything we can!

Read the above quote again: it starts with the phrase that theirs is a 'quaint' hope—meaning that their desire is that readers will understand and believe that it is written with no other objective than that which is stated—an honest attempt to reduce poverty. Next, move on to the sentence that baldly but simply states the reality of the situation: the United States—all of us together—are running a "high-poverty economy" with little thought to those "externalities" like hunger, childhood trauma, deep poverty, homelessness, and lack of education that cause extreme suffering, and cost society boatloads of money. Then move on to the next phrase—"authentically focused"—meaning that so many so-called reform efforts are *not* authentically focused on the real objective of truly *reducing* the poverty rate, but playing instead the political game. Finally, their use of the word "legitimate" indicates the intent to keep poverty reduction programs in line with the facts and evidence, and also with the human caring heart—not the current political ethos. All of this makes you just sit there because there is nothing else to do at the moment—but just for the moment.

As we begin to further explore this situation, a brief explanation about *poverty measurement* is necessary if we are to truly comprehend *how* such substantial poverty endures in the richest country in the world. First of all, the United States continues to use an admittedly antiquated measurement process developed in the 1960s. This measurement known as the "Official Poverty Measurement" (OPM) remains in official use however. When the US began to measure poverty for the first time as a result of LBJ's 1960s "War on Poverty," income—as measured by food insufficiency—became the demarcation of the "poverty line," non-poor above and the poor below. This OPM, while now recognized as clearly inadequate, still remains in use: food costs have gone down. But the not-included costs of housing, medical, and childcare have gone up; however, they are not included in the official measure. Evidently, the OPM's simplicity remains an advantage.

But two other measures—"Supplemental Poverty Measure" (SPM) developed in 2009 and the "Survey of Income and Program Participation" (SIPP)—enable more precise measurement. This greater precision of measurement is, however, raising increasing number of questions about how many people suffer from poverty. For example, the Public Policy Institute of California, working in conjunction with the Stanford Center on Equality and Inequality, found that while the Official Poverty Measurement (OPM) defined 16% of California families living in poverty in 2011, the SPM measured poverty at 23.5% for that same year. There are now multiple challenges to the OPM measurement and its capacity to reflect the full scale of poverty in the United States and challenges to the SPM as well. Hopefully, a better process will be selected—we shall wait and see.

A brief history of *official poverty reduction initiatives* is also necessary if we are to have "an open and honest discussion of the facts," as the Stanford Center authors request. The "Aid to Families with Dependent Children" (AFDC)—a federal cash assistance program assisting poor mothers and dependent children—was a much criticized but still steady support system for 60 years to people in need. It was replaced in 1996, by the Clinton Administration working in concert with the Republican Congress. Together, they initiated the "Personal Responsibility and Work Opportunity Reconciliation Act" (PRWORA).

This Act consists of two programs: first, "Temporary Assistance to Needy Families" (TANF), a federal block grant to States designed to increase employment among low-income populations, but at the same time remove them from the federal cash assistance program after a prescribed amount of time. And second, an expanded "Earned Income Tax Credit" Program (EITC), managed by the IRS, which was designed to shift the safety net toward employed parents. While EITC has grown exponentially ($62 billion in 2018), government spending on basic cash assistance to people in poverty has declined from about $21 billion in 1997 to about $8 billion in 2018.[8] Temporary expansion of EITC to aid children in poverty was expanded during the Covid-19 years and is currently, in 2024, the subject of much political discussion.

Despite some short-term success of this TANF program—two to three years in the late 1990s—the longer-term outcome of the current work-program safety nets is not at all impressive; EITC also needs some redesign. Although they both reduce poverty for those close to the non-poverty line, neither is able to keep poverty from increasing in more vulnerable groups. It is useful to note that this increasing discernment of real-time impact was partially due to the supplement measures (SPM and SIPP) mentioned above.

As a result of these assessments, according to the Stanford researchers, "there is a *growing misalignment* … between the key features of our safety net and the characteristics and circumstances of the contemporary low-income population." Two key reasons the author-researchers identify for this mismatch are the "rise in jobless poverty" and the "rise in childless poverty"—they label these as "stylized facts" which inform and impact current proposals to reduce poverty.[9]

Stylized facts are simply a constellation of facts shaped to fit a preconceived belief structure. But Philip Alston, in that previous report, identifies—based on fact and evidence—seven specific components necessary to eliminate poverty. They are (i) democratic decision-making; (ii) full employment policies; (iii) social protection for the vulnerable; (iv) a fair and effective justice system; (v) gender and racial equality and respect for human dignity; (vi) responsible fiscal policies; and (vii) environmental justice.

Alston observed in 2017 that the United States fell short on each of these issues. At that moment, many avidly argued against his points. However, I believe that there are far fewer who would even attempt to argue against Alston's point today, given recent data research on what actually works to reduce poverty, and the US experience with Covid-19 as well.

Perhaps the best example underscoring Philip Alston's point that the persistence of poverty in America is a political choice, is Mathew Desmond's newest book, *Poverty by America*. As Desmond illustrates, we are the richest country in the world, and yet we have more poverty than any other wealthy country. He asks why. Given the circumstances, he focuses on new anti-poverty actions to consider. But his conclusion tells it all: "*we must become poverty abolitionists … refusing to live as unwitting enemies of the poor.*"[10]

Political Undertow and Its Mythologies

Our two political parties define the problem of poverty in two fundamentally different ways. Those who claim "welfare dependency" is the real problem believe that if people would just get a job, learn to work hard, and get married before having children, poverty would substantially decrease. Those who claim "inadequate resources" is the critical problem believe that necessary basic cash assistance and implementation of more effective programs will extricate people from their dire situations.

Both of these political belief systems have negative thought-world foundations based on the supposed reality of scarcity and how to overcome it. The

Republican welfare dependency perspective is shaped by down-deep belief in "negative difference": if the person can't work hard, it's her/his/their fault, and they deserve no assistance. The Democrats' "inadequate resources" perspective is a bit more complex. By inserting government help into individual lives, it first makes Democrats an easy target for Republican charges of socialism and communism which, according to this resulting mythology, denies the primacy of the individual. But the Democrats make it worse for themselves by not insisting strongly and clearly that assistance to the individual can and should be expanded, funded by the government, and given to the best organizations available to offer reciprocal care.

Poverty experts tell us, however, that this particular political 'tug-of-war' resolved itself more than two decades ago in favor of those who believe in "welfare dependency and negative difference," and they're sticking with it, no matter the evidence.[11] *Pathways,* a publication of the Stanford Center on Poverty and Equality, is a useful knowledge guide offering theory, data, and practice information from all sides.[12] It is a web and hard copy magazine which reports on poverty and inequality trends, cutting edge research, key interventions, and the newest vision and debates among commentators. As such, it gives us the real-world answers.

Of course, it is always advisable for in-depth understanding, to also read books about the effects of poverty on both the person and the community. The chronicling of severe trauma by Mathew Desmond in *Evicted: Poverty and Profit In the American City*; or David Shipler's *The Working Poor: Invisible in America* give necessary background because they put heart as well as experiential evidence into the data. Otherwise, we cannot begin to understand. As Shipler tells us: "most of the people who I write about do not have the luxury of rage. They are caught in exhausting struggles."[13]

The Stanford *Pathways* information and analysis offers a different perspective that is particularly important if we wish to eradicate poverty. Their information allows a view into *how* poverty endures. The winter 2016 issue, "The Poverty and Inequality Election" illustrates and analyzes the political differences which drive the very different policy interpretations guiding our national policy interventions on poverty. It begins this analysis with two articles that define how Republicans and Democrats believe poverty can be reduced. Without first comprehending the differences which make up this ongoing tug-of-war and difficult political undertow, it is almost impossible to understand either the reality or the implications of US poverty policy action today.

"Reducing Poverty the Republican Way" defines four principles that "conservatives and Republicans should use to guide the agenda for the future." The first principle is to make sure to solve the right problem according to the "Republican Way," and the problem is definitely *not* poverty. Instead, the "problem is that too many Americans are not self-sufficient." As a result, the second principle states, "all policies should be pro-work," as that is "the dividing line between the poor and non-poor." The next principle, predictably, is that "taxpayer dollars must be accompanied by accountability for outcomes," while the fourth and final principle summarizes a long-held Republican belief which necessarily shapes the policy: "Federal programs will fail without a social foundation of better parents and stronger marriages."[14]

"Reducing Poverty the Democratic Way" takes into account Democratic economic and social concerns as well as those "consistent with the broad values of Americans." This diagnosis first leads directly to the prescription of policies that will "raise education and skills among poor children, youth, and adults." Second, living wages are a necessity for all, even the unskilled; third, "full employment," with job offerings open to all, and there is a necessity to "address the specific problems of such groups as ex-offenders, noncustodial parents, children in very poor families or neighborhoods, and people with disabilities."[15]

Given these two very different political approaches to poverty, it's also useful to understand the differences between Democrat and Republican *voters* themselves on these issues. Pew Surveys, in December 2015, interviewed both Democrats and Republicans on the issue of poverty in all three income categories: low-income category (below $30,000); middle-income ($30,000 to $74,999); and high income ($75,000+). The question asked was if the "federal government should play a major role in helping people get out of poverty." Averaged together: "Democrats were nearly 35% more likely than Republicans to say poverty reduction 'should play a major role' in federal policy." This response was no surprise.

But the most revealing response was a surprise—and it was the differences *within* parties. Question differences between low- and high-income Democrats were a mere 10%: 78% of low-income participants and 68% of high-income participants approved of their Party's approach to poverty eradication—nothing surprising there. However, it was the relatively large difference *within* the Republican Party that was the big surprise. While 53% of low-income Republicans support the Democrat federal poverty reduction policy, only 24% of high-income Republicans do. That is a difference of almost 30%. A 2014 Pew survey about equality/inequality also indicates somewhat similar differences within the Parties. Democrats had a 90% rate for government intervention to counteract inequality, while Republicans were evenly divided.[16]

Overall, these outcomes simply underscore long recognized differences between the parties: Republicans, particularly those that are doing well, prefer to rely on markets and the necessity of individual work, while Democrats are more willing to support a bigger government in order to achieve their goals of poverty reduction. As a result candidates of both parties, for both 2016 and 2020, had a variety of proposals, ranging from simple voicings of intent to fully costed-out proposals—but always based on their party's ideological commitments.

As a result, two well-known institutions, Brookings and the American Enterprise Institute—one center-left and one center-right—collaborated on a report which argued "that both the causes and solutions of poverty and opportunity fell into clusters pertaining to *"family, work and wages, and education."* The hope was that both parties would adopt a strategy that "would mount simultaneous attacks in all three domains."[17] So far, that has not happened. And it all goes back to hewing to those "stylized facts," especially for Republicans, rather than looking at the reality of facts, evidence, and consequences.

But in reality, there also seems to be more to it than that—there is something else that is still missing. Clearly, first understanding the political postures of each party—in their own words—and then examining the perspectives of voters about these party postures tells us something is off, something doesn't quite fit reality. It indicates a fuzzy but real gap in understanding how poverty is conceived by the political parties, perceived by the voters, and then designed and acted upon by poverty officials—but it is not clear at first glance. It makes sense then to understand what this gap is about. We do so by pursuing poverty facts and evidence to clarify what's actually missing in our understanding.

Debates Behind the Strategies

These "stylized facts" are critical. Both programs in the 1996 PRWORA Act focus on employment and work, which are exactly contrary to the emerging real facts. TANF programs offer time-regulated limited resources and training while mothers with children attempt to find work. But after an established period of time, regardless of their situation, they are forced out of the program—and then what? At the same time, EITC offers expanded wage subsidies *only* to working families with children living at home—i.e., custodial children (although there are now efforts to expand this program to those without "custodial children"). This means that adults—single or married—who are suffering from poverty do not qualify for assistance because they have no children.

At the same time over the past six decades with fewer jobs available, even with episodic upticks, there has been a well-documented decline in the employment rate of prime-age men. This general trend, while still somewhat cyclical, "has been relentlessly downward." Women's employment, due to changes in gender equality, increased in prime-age employment until 2000, "but thereafter the decline has also been steep"; but again, with episodic upticks.[18] These trends, which increase both "childless poverty" and "jobless poverty," relentlessly expand overall poverty. Currently, after Covid there is a substantial upswing, but its sustainability remains in question.

The debate among politicians as well as among scholars concerning the reasons for this decline in employment has been described as "contentious." Some scholars labeled the decline as "reductions in labor supply"—in other words, "'choosing' not to work"; while others emphasized "reductions in the demand for labor" (particularly demand for low-skill labor) as the cause, meaning a "decline jobs available." Thus, we can see how resulting divisions remain political. However, after a number of years of ongoing debate, the bi-partisan Council of Economic Advisors (CEA) issued a report in 2016 concluding that *"reductions in labor supply—are far less important than reductions in labor demand* in accounting for the long-run trend."[19] In other words, it's the lack of jobs—not willingness to work—that is the problem.

But the poverty researcher authors—ever conscious of putting all the facts on the table—caution that these findings do not mean that "labor-supply effects (choosing not to work) are *entirely* irrelevant" (emphasis theirs). They observe that a small but growing group of adults are indeed disconnected from work. Their unresponsiveness is, for the most part, attributed to one of the following: (i) rising use of disability insurance; (ii) growing geographic immobility; or (iii) rising incarceration rates with resulting criminal records blocking re-entry to the labor market.[20] Of course, in 2024 we are still emerging from the Covid years. Labor demand is surging, with inadequate immigration rules playing a role. At this same point in time, deep poverty is also surging. Sadly, however, these are the kind of meaningful facts which are hard to include in a "gotcha" political debate currently primarily manufactured by the Excluder hard-right.

The second highly stylized fact affecting poverty reduction efforts, "*Childless poverty*" is particularly harmful and debilitating because it most often comes arm-in-arm with the previously discussed "*Jobless poverty.*" Childless poverty refers to poor adults who have no children or who are not living with their children. The consequent fact is that with no custodial children, these adults have *no access* to national poverty programs. This second fact also strongly shapes the structure of poverty in the United States. The poverty researchers state: "The rise of childless poverty emerged gradually over the past 40 years. Among poor adults who are 18 to 64 years old, 47 percent were childless in 1975, while 58 percent were in 2015."[21] I do ask myself if this recent shift in poverty could account for the increasing numbers of homeless encampments that we see in cities? Perhaps, if we analyzed poverty according to the real facts rather than the stylized facts, we might have greater positive impact on both people and places.

Three main societal forces are behind this slow but consequential rise: (i) the waning baby boom; (ii) the increase of delayed marriage; and (iii) the emergence of complicated families. Childless poverty has increased for all ages, but particularly for those under the age of 50. For poor adults, *ages 35 to 49 years*, poverty increased from 22% in 1975 to 43% in 2004. In comparison, for poor adults, *ages 18–34*, childless poverty was 46% in 1994 but climbed to 61% in 2015. The researchers note that similar rates of "non-custodial status" are also found among the *non-poor* of similar ages ranges, clarifying that the poor population is simply reflecting change taking place in the entire population.[22]

As a result, *childless poverty* is a second example of the misalignment between our current poverty reduction programs and the reality of the poverty population. With rates of employment foundering and traditional two-parent working families no longer the norm, it clearly brings into question the viability of the present focus of EITC with its requirements of *work* and *children* as the two pre-requisites for enrollment.

One of the more disturbing implications of this mismatched and misaligned poverty policy is the rise of "deep poverty" in terms of "disconnected families and the emergence of $2-a-day poverty. Even by 2000 randomized experiments with poverty programs indicated a troubling new stratification of the poor. While the 1996 initiated PRWORA programs reduced poverty overall with their focus on the less-poor working families—these same programs actually increased deep poverty. For example, the SNAP (Supplemental Nutrition Assistance Program) reported 289,000 SNAP households with children having no cash income in 1995. But by 2015, this number had grown to just under 1.3 million households according to those families reporting zero cash income at the SNAP office. (All SNAP recipients are required to report annual incomes under penalty of law.) The poverty researchers conclude that "rather than roughly doubling since welfare reform, $2-a-day poverty tripled or quadrupled."[23]

All of these facts, and all of this evidence, return us to that implicit message given at the beginning by the author-researchers in "Facts Behind the Vision." I initially characterized their message as "Do Something." The facts and evidence clearly suggest that the vision of politicians—particularly those that presently insist, according to the researchers, on solving poverty the "Republican Way" as earlier presented, are definitively missing the mark. The programs create too much suffering for too many children and adults. But even more troubling, it is also the "tepid response" by we citizens—in reality, poverty doesn't seem to be of great interest to us.

This might be because we cling to that popular concept of "mobility optimism." It has been a long-enduring feature of American life—featuring competition open to all. As a result, it is assumed to give everyone the same chance. But in reality, the real-time present data now indicates, instead, that the impact of positive "birth lottery"—being born into a high-earning family—is the critical factor in "getting ahead" in the United States. Further, the same data indicates that upward mobility is now substantially higher in other countries than in the United States and none of these come with the moniker "the land of opportunity."[24] So, yes, sadly "mobility optimism" is no longer a working feature of our communities.

But maybe we don't pay that much attention to poverty and tepidly cling to mobility optimism for another reason. In all honesty, the poverty problem may seem just too big, the difficulties too well entrenched—as perhaps do all the four ancient wrongs. And as a result, what we characterize as our individual, or small grassroots effort, just does not seem worth the effort.

On the other hand, so far we have not looked beyond the supposed realities of our everyday life. In fact, what may be missing from our understanding is that the answers are right there, if we choose to listen. If we first begin to question those old negative thought-worlds and the beliefs and attitudes that come with our everyday lives, and instead, look around us at reality, it may be possible to discern something quite different. The Stanford Center on Poverty and Inequality present different and compelling suggestions for us all that actually expand beyond our everyday reality.

They begin with a comment: "It is perhaps a puzzle that a country so rich is seemingly so untroubled by poverty." The Stanford groups then spells out two options. The first is that we continue with what is defined as *Conversation #1*: "narrow solutions to overcome tightly defined problems"—sometimes summed up as the "nudge" philosophy. This is essentially what we are already doing, but paying little attention to. The poverty researchers then suggest something quite different as *Conversation #2*. "In addition to any expedient, small-scale interventions, we have a wider discussion about where poverty comes from and what types of larger-scale changes might be needed to eradicate it."[25]

The author-researchers then immediately note that this is a "difficult conversation" to have. With some long-term experience in this area, I couldn't agree more—it is difficult. For example, in my own personal experience, in every work meeting that I have ever attended about poverty—and there were many—I would often ask the question: "can we say 'eradicate poverty' instead of just repeating 'reduce poverty'?" This was just another way of asking: "can we have 'Conversation #2' as well as Conversation #1'?" I admit, I rarely got the answer I wanted—most resembled my uncle's response to me as a young child—"that's just the way the world is." This negative overall philosophy has permeated poverty action, and the four ancient wrongs as well, for far too long. The poverty researchers recognize it, and they are finally asking us to do more. They ask us to look at the reality of the situation and not just the promotion of political context.

The Devastated and Exploited

The poverty researchers at the Stanford Center on Poverty and Inequality have given us important information on *how* poverty reduction programs work—and don't work. But they also communicate—again and again—a much more person-centered message that centers on the issue of stigma and rejection. One Stanford group described it this way:

> To be a welfare recipient was to wear a scarlet letter in the eyes of fellow Americans, one that robbed you of dignity and self-worth. Due to the stigma, the program served to isolate the poor from—rather than integrate them into—the rest of society. We have then, an implicit social contract

with the poor, a social contract that implies that anyone who uses the available relief is effectively disenfranchised.[26]

In a similar fashion, the leader of the new version of the "Poor People's Campaign: A National Call for Moral Revival," the Rev. Dr. William J. Barber II, tells us in speaking of America, that we will "never even get close to being a more perfect union—until we are honest about her past and the politics of rejection." He then explains how "the Christian faith has been used to whitewash the rejection that stains our nation's soul."

That is the soulful preacher talking to us, and we need to hear it, but Rev. Barber is also the pragmatic and peace-making revolutionary when he tells us "the *politics of rejection* and *policy violence* against the poor are still far too real" (emphasis added). For example, 1.5 million public school students were reported to be homeless during some period of the 2017–2018 school year, and now, there are many more. Reports indicate that LGBTQ+ youth are at an even higher risk for homelessness. So, he calls on the poor—the rejected—to come together, to be a movement for love and justice. So my question is: can we, in support, join them?

Does that sound too idealistic and aspirational? It's not really. It is the poor and the rejected that can tell others, with some level of consummate expertise the effects of the policy violence and rejection, and how to begin deconstructing that violence as well as those political lies which obscure the reality that is right in front of us—if we would only look. Four groups of rejected peoples, whose disenfranchisement has clearly been created with distinctive levels of policy violence, particularly since the 1970s, can be recognized as both participants and participatory leaders of this Movement.

This is the same movement which Martin Luther King began more than 50 years ago in 1968. These four groups, similarly identified by the Stanford poverty researchers, are the following: (i) those poor who have been consigned to the incalculable suffering of deep poverty; (ii) poor immigrants who fear government authority and coercion regardless of their legal status; (iii) the formerly incarcerated who are blocked from re-entry to a normal work-life because of criminal records; and (iv) minimum-wage and gig-workers, suffering from anxiety every day due to the unnecessary imposition of scarcity by economic system.

Deep and extreme poverty has grown substantially in the past two decades as increasing numbers of poor parents with children have been unable to find jobs, and then basically excommunicated from any further financial assistance by TANF rules. In 1999, 40% of the poverty population was living in deep poverty, defined as "income less than half of the poverty threshold," whereas by 2015 the percentage had risen to 46%. But there is an even deeper level of poverty—those living on less than $2.00 a day—labeled "extreme poverty." These extremes began to form as a result of TANF policies that pushed mothers off of welfare rolls after a specified amount of time if no job was forthcoming.

In the book *$2 a Day: Living on Almost Nothing in America,* the authors state that in 2016 at least one and a half million households, and more than three million children found themselves in this situation. And yet to be documented are those now pushed there by Covid. But let's be clear: both levels of poverty—deep and extreme—have been set in motion by the design of the national US poverty program, TANF, which we can and should now recognize, I believe, as extreme "policy violence." It makes finding a job an absolute necessity when there are often fewer jobs; and when there are many, they are mainly gig jobs with no security whatsoever. Is this "the land of opportunity's" idea of a grotesque joke? And while the minority who actually do find a job, are labeled the "deserving" poor; the rest, exhausted and devastated, are obviously if implicitly labeled the "undeserving" poor.[27]

Poor immigrants (separate from asylum seekers) contributed to major changes in the makeup of the poverty population in the past two decades. After 1999, while white and black representation in this population diminished, Latinx representation expanded; they now represent 18% of this population group.[28] From a diversity standpoint, this has implications for poverty program effectiveness in terms of language learning and changing neighborhood segregation patterns. But the biggest problems are in terms of the reach and use of safety net programs.

Although the great majority of Latinx people are either citizens or green-card holders, all are caught in widening worries about deportation given the rate of extreme *policy violence* which was mounted over the past decades with a brutal implementation of immediate deportation—no questions asked—practiced with impunity. These policy violence practices include family separation policies, not only at the border but also for the deportation of parents living in the United States; immigration raids on businesses by ICE—arresting only workers but not owners and managers in these raids. Illegal asylum rules: those that put children in cages, and illegal arrests of adults as they cross the border—"no crime" according to accepted international law—multiply annually. All of this contributes to people and children living in fear, unable to access even the basic necessities.[29]

Formerly *incarcerated peoples*, having paid their debt to society in prison, find themselves pushed aside, and their rights as citizens violated. Lani Guinier, in discussing the now celebrated book *The New Jim Crow* by Michelle Alexander, characterizes it best: "dreary felon garb, post-prison joblessness and loss of voting rights now do the stigmatizing work done by colored-only water fountains and legally segregated schools."[30] That juxtaposition of ideas certainly makes us begin to think differently, doesn't it?

The extreme and violent policy violations are multiple. The United States has the highest incarceration rate in the world, almost five times more than the average for other wealthy countries—and two-thirds of those incarcerated are people of color. The total *female* jail population increased from 8,000 in 1970 to nearly 110,000 in 2014, but more than 80% of these women are imprisoned for non-violent offenses. "By the Department of Justice's own admission 95 percent of the growth in the incarcerated population since 2000 is the result of an increase in the numbers of un-convicted defendants, many of whom are unable to make bail!" What? Did we hear that right? And yet, for those that have completed their imprisonment, there is one more policy violence—and rejection—to be endured; no job because of a criminal record, a final and lasting indignity.[31]

Minimum wage workers, even though working long and hard hours, do not have a pathway out of poverty. "Gig-work" also does not provide either the wage or the hours for any kind of real security. In addition, researchers have found that hourly wages for low-skill male workers have actually declined, and substantially so, between 1980 and 2012. For example, real hourly earnings fell between 1980 and 2012—22% among high school dropouts and 11% among high school graduates.[32] As a result, there are now 64 million people working for less than $15 an hour.[33] There is one more mind-blowing fact; this is not illegal because the Federal Minimum wage remains at $7.25. Numerous city and local community boards, recognizing the need for a raise in minimum wage, have attempted to do just that. Once these new rules passed at the community level, legislatures in 27 states have then passed new laws *nullifying* the community-passed raises in minimum wage.[34]

Sprinting across these terrible injustices has its own terrible dangers. The lack of sufficient attention to the incarceration of black people is one of those. In addition to *The New Jim Crow*, Brian Stephenson's book *Just Mercy* defines the particular death penalty reality and its injustices that should be truly known and recognized. The film of the same name helps to expand that awareness. At the end of the film, Stephenson sums it up unconditionally. He tells us: "the opposite of poverty is justice." Stephenson begins his book with a quote from Reinhold Niebuhr: "Love is the motive, but justice is the instrument." He concludes his book by telling us "Mr. Hinton became the 132nd person in America exonerated and proved innocent after having been wrongfully convicted and sentenced to death".... Stephenson also tells us: "In the wake of the 2010 Supreme Court ruling banning life imprisonment without parole for children convicted of non-homicide crimes, hundreds of children condemned to die in prison are now being resentenced and dozens have already been released."[35] Yes love, caring, and justice, are the instruments.

Finally, let's not forget the soulless and interminable rejection of eviction and homelessness. *Evicted: Poverty and Profit in the American City* by Mathew Desmond is the best book I have ever read on this issue. Actually, that's not really true; I have never totally finished reading the book. I read as much as I can, and then put it aside, promising myself to go back to it so

I can complete the chapter. But down deep I know I never will—it cuts too close to home.

When I was 14, my Dad lost his job—a good one that had allowed us a comfortable middle-class life with a pretty house, my own bedroom, and lots of friends. And then everything changed. My mother, my two brothers, and my sister—we all followed my Dad to another State where he thought he had another good job. But he didn't, so we ended up living in a "motel with kitchenette." One day, walking to the grocery store, I saw a pretty, white clapboard house with a "for-rent" sign. When I went into the grocery store, I asked the cashier—it was a small town—who was the owner of that house, and where did they live?

She told me "up the hill" and I started out. When I arrived at Mrs. S's house—I still remember their name—I quickly knocked on the door, and the owner opened it with a questioning smile. Long story, short: Mrs. S. invited me in and offered me a coke, while I told her why we wanted to rent her house. We sat there, and Mrs. S. said that she would have to speak to my mother. My mother, understanding the need and the anxiety behind my actions, went to speak with Mrs. S.—she was invited in for coffee, and we had a house once again—only because, I now realize, the owners made it easy for us—and let's just put it out there—everyone was white (although we were also Irish Catholic). But even so, we were some of the lucky ones: although we suffered the anxiety and instability of having no house for a time, we never suffered eviction, and we were able to get back on track. But if that small trauma stays with me after all these years, what does actual eviction do? Desmond endeavors to tell us.

> *Evicted* follow eight families—some black, some white; some with children, some without—swept up in the process of eviction ... Evictions' fallout is severe. Losing a home sends families to shelters, abandoned houses, and the street. It invites depression and illness, compels families to move into degraded housing in dangerous neighborhoods, uproots communities, and harms children. Eviction reveals people's vulnerability and desperation, as well as their ingenuity and guts.[36]

Desmond's book attests to the numbers of the evicted: "Every year in this country people are evicted from their homes, not by the tens of thousands or even the hundreds of thousands, but by the millions." Desmond tells us: "Until recently we simply didn't know how immense this problem was, or how serious the consequences, unless we had suffered themselves ourselves. For years, social scientists, journalists, and policymakers all but ignored eviction, making it one of the least studied processes affecting the lives of poor families." He goes on to say: "But new data and methods allowed us to measure the prevalence of eviction and document its effects."[37] Even so, we all know that the next stop on the road after eviction, for far too many, is the final dislocation and rejection of homelessness.

Rejecting Indifference

In 2017—pre-Covid-19—rejection, policy violence, and indifference clearly prevailed for the poor. Let's briefly summarize the facts: 52.1% of children under the age of 18 were poor or low-income (38.5 million children). Even the youngest and most vulnerable do not fare well: data indicates that "due to consistent underfunding, Early Head Start served only 5 percent of eligible infants and toddlers in 2016." Finally, using the more precise poverty measurements spoken of earlier, there are *140 million* poor and low-income people in the United States. To make the inequality point, just three very wealthy individuals in the United States possess a combined wealth of *$248.5 billion* in 2017. According to Forbes magazine "this is a level and amount of wealth among these three billionaires that is equal to the bottom 50% of the country."

Something is wrong in this country. In a supposed democracy—where duly elected legislators represent the will of the people—the multitude of acts creating these statistics have to be the ultimate sanctification of indifference—not just by the wealthy and the rich, but by all of us as members of this democracy. In 2020—mid-Covid—the devastation of policy violence and the politics of rejection was made much clearer—and the indifference which had been so relentlessly expanded and utilized for profit, particularly over the past 50 years, began to haunt our entire nation—as we witnessed on TV, in the comfort of our homes, the parade of cars lined up for free food.

In my own heart, the situation we now find ourselves in is somehow encapsulated by the life of Honestie Hodges. I read a newspaper article about Honestie in 2017. As an 11-year-old girl, she was brutally handcuffed by police outside of her home in Grand Rapids, Michigan, for no reason. The police chief later said that he was 'nauseated' by the video, but he did not fault the police officers as they had followed police "procedures" in a case of mistaken identity.

Honestie's face, and her mother's reported screams, "no! She is eleven years old, sir" stayed in my dreams and thoughts for a long time. Recently, I read that Honestie had died of complications from Covid-19 on Sunday, November 22, 2020, just several days after her 14th birthday. Days before Honestie's death, her grandmother had posted a "GoFundMe" request for financial assistance for an imminent operation possibly needed by Honestie because of Covid-19 complications—but that never happened.

Several days before Honestie's death, on November 19, 2020, the American Medical Association (AMA) the nation's largest physician association, declared "racism a public health threat": "As long as there are health inequities, 'the overall health of the nation will suffer' the AMA emphasized." This statement was a follow-up to the AMA's June 2020 "pledge to confront systemic racism and police brutality." Kudos to them: but it also seems to be a hollow follow-up to the vast inequities of Honestie's life with its clear intertwining of racism, poverty, and health inequities. Somehow, for me, Honestie's life brings together all of the vile brutality that we have let embroil those consigned to poverty or living on the margins, to live within. And all of that continues to go on without sufficient contestation in these United States of America.

Yes, "DO SOMETHING" is the right message.[38] And we can indeed, do a great deal if we so wish. Remember, Philip Alston told us the components necessary to eliminate poverty: (i) democratic decision-making; (ii) full employment policies; (iii) social protection for the vulnerable; (iv) a fair and effective justice system; (v) gender and racial equality and respect for human dignity; (vi) responsible fiscal policies; and (vii) environmental justice. These are the categories from which you and your deep participation group can make decisions of specific reforms and reinventions to accomplish. So, yes. Let's do it.

Notes

1 OECD. The United Nations Office of Economic Cooperation and Development origins date back to 1960, when 18 European countries plus the United States and Canada joined forces to create an organization dedicated to economic development. Today, 36 member countries span the globe, from North and South America to Europe and Asia-Pacific. OECD.org has numerous excellent reports on poverty and inequality online.
2 Philip Alston, "Report of the Special Rapporteur on Extreme Poverty and Human Rights on his mission to the United States of America", United Nations, 2017, https://digitallibrary.un.org.
3 Rev. Dr William J. Barber II, *We Are Called to Be a Movement* (New York: Workman, 2020), 37.
4 Alston, "OECD Poverty Report", paragraphs 10, 11.
5 Philip Alston, "OECD Poverty Report", 2017, 2.
6 Charles Varner, Marybeth Mattingly, and David Grusky, "The Facts Behind the Vision", *PATHWAYS: A Magazine on Poverty, Inequality, and Social Policy*, 2016, 3–8, https://inequality.stanford.edu
7 Varner et al, "The Facts Behind the Vision", 3. https://inequality.stanford.edu
8 Gordon Berlin, "The Tug of War Between Worrying About Poverty and Worrying About Dependency", *Pathways*, Winter 2018, 29. https://inequality.stanford.edu
9 Varner et al, "The Facts Behind the Vision", 3 https://inequality.stanford.edu
10 Mathew Desmond, *Poverty by America* (New York: Crown, 2023).
11 Gordon Berlin, Winter 2018, "The Tug of War", *Pathways*, 29. https://inequality.stanford.edu
12 *Pathways Magazine*, The Stanford Center on Poverty and Inequality publishes their journal "PATHWAYS" several times per year. I have used the publications from 2016 through 2019 in this chapter because they explore all sides. For people interested in these subjects, I encourage utilization of these research-based

publications. They are truly excellent and are available free online, or as hardcopy with free sign-up
13 Mathew Desmond, *Evicted: Poverty and Profit In the American City* (New York: Broadway Books, 2016). David K. Shipler, *The Working Poor: Invisible in America* (New York: Knopf, 2004).
14 Douglas Holtz-Eakin, "Reducing Poverty the Republican", *Pathways*, Winter 2016, 15.
15 Harry J. Holzer, "Reducing Poverty the Democratic Way", *Pathways*, Winter 2016, 19.
16 Ron Haskins, "What Are the Potential Candidates Saying About Poverty and Opportunity?", *Pathways*, Winter 2016, 4
17 Ron Haskins, "What Are Potential Candidates Saying", *Pathways*, Winter 2016, 5. https://inequality.stanford.edu
18 Varner et al, "The Facts Behind the Vision", *Pathways* 4. https://inequality.stanford.edu
19 Varner et al, "The Facts Behind the Vision", 4. https://inequality.stanford.edu
20 Varner et al, "The Facts Behind the Vision", 4. https://inequality.stanford.edu
21 Varner et al, "The Facts Behind the Vision", 5. https://inequality.stanford.edu
22 Varner et al, "The Facts Behind the Vision", 5. https://inequality.stanford.edu
23 Luke Shaefer and Kathryn Eden, "Welfare Reform and the Families it Left Behind", *Pathways*, Winter 2018, 24–26. https://inequality.stanford.edu
24 David B. Grusky, Marybeth J. Mattingly, and Charles Varner, "The Poverty and Inequality Report: Executive Summary", *Pathways*, 6. https://inequality.stanford.edu
25 Michelle Jackson, "Don't Let Conversation One Squeeze Out Conversation Two", *Pathways*. Spring 2017, 34. https://inequality.stanford.edu
26 Kathryn Edin, H. Luke Shaefer, and Laura Tach, "A New Anti-Poverty Litmus Test", *Pathways*, Spring 2017, 10. https://inequality.stanford.edu
27 Varner et al, "the Facts Behind the Vision, 4. https://inequality.stanford.edu
28 Varner et al, "the Facts Behind the Vision, 7. https://inequality.stanford.edu
29 Francisca Antman, "Contemporary Social Issues and the Latinx Experience in the US", Webinar, Economic Policy Institute, Dec. 2, 2020. https://epi.org
30 Lani Guinier, book jacket of *The New Jim Crow* by Michelle Alexander (New York: New Press, 2012).
31 Poor People's Campaign, "A Moral Agenda Based on Fundamental Rights", Washington D.C. p.7. https://poorpeoplescampaign.org
32 Varner et al, "the Facts Behind the Vision", *Pathways*, 6. https://inequality.stanford.edu
33 Poor People's Campaign, "A Moral Agenda Based on Fundamental Rights", Washington D.C. 3. https://poorpeoplescampaign.org
34 Poor People's Campaign, "Fact Sheet on Children", https://poorpeoplescampaign.org
35 Bryan Stevenson, *Just Mercy: A Story of Justice and Redemption* (New York: Spiegel and Grau, 2014), 315–316.
36 Mathew Desmond, *Evicted: Poverty and Profit in The American City* (New York: Penguin, 2016), 5.
37 Mathew Desmond, *Evicted*, 295, 296.
38 Shani Saxon, "Honestie Hodges, Whose Handcuffing Sparked Outrage, Dies of Covid-19 at 14", *ColorLines, https://colorlines.com*. This is good source of news for race, power, and democracy.
Kevin B. O'Reilly, "AMA: Racism is a Public Health Threat", Nov. 2020, https://ama-assn.org. This is a good article to read and see if the AMA is actually following up and doing what they promise in this article.

Bibliography

Alston, Philip. "Report of the Special Rapporteur on Extreme Poverty and Human Rights on his mission to the United States of America", *OECD*, United Nations, 2017.

Antman, Francisca. "Contemporary Social Issues and the Latinx Experience in the US", Webinar, Economic Policy Institute, Dec. 2, 2020.

Barber II, Rev. William J. *We Are Called to Be a Movement*. New York: Workman, 2020.

Berlin, Gordon. "The Tug of War Between Worrying About Poverty and Worrying About Dependency", *Pathways*, Winter 2018.

Desmond, Mathew. *Poverty by America*. New York: Crown, 2023.

Edin, Kathryn, H. Luke Shaefer, and Laura Tach, "A New Anti-Poverty Litmus Test", *Pathways, A Magazine on Poverty, Inequality, and Social Policy*, Spring 2017.

Grusky, David B., Marybeth J. Mattingly, and Charles Varner, "The Poverty and Inequality Report: Executive Summary", *Pathways, A Magazine on Poverty, Inequality, and Social Policy*, 2016.

Guinier Lani. "Book Jacket", *The New Jim Crow* by Michelle Alexander. New York: New Press, 2012.

Haskins, Ron. "What Are Potential Candidates Saying", *Pathways, A Magazine on Poverty, Inequality, and Social Policy* Winter 2016.

Douglas Holtz-Eakin, "Reducing Poverty the Republican Way", *Pathways, A Magazine on Poverty, Inequality, and Social Policy*, Winter 2016, 15

Holzer, Harry J. "Reducing Poverty the Democratic Way", *Pathways, A Magazine on Poverty, Inequality, and Social Policy*, Winter 2016.

Jackson, Michelle. "Don't Let Conversation One Squeeze Out Conversation Two", *Pathways, A Magazine on Poverty, Inequality, and Social Policy*, Spring 2017.

Poor People's Campaign, "A Moral Agenda Based on Fundamental Rights", Washington D.C, 2022.

Saxon, Shani. "Honestie Hodges, Whose Handcuffing Sparked Outrage, Dies of Covid-19 at 14", *ColorLines*, Nov. 2020.

Shaefer, Luke and Kathryn Eden. "Welfare Reform and the Families it Left Behind", *Pathways, A Magazine on Poverty, Inequality, and Social Policy*, Winter 2018.

Stevenson, Bryan. *Just Mercy: A Story of Justice and Redemption*. New York: Spiegel and Grau, 2014.

Varner, Charles, Marybeth Mattingly, and David Grusky, "The Facts Behind the Vision", *Pathways, A Magazine on Poverty, Inequality, and Social Policy*, 2016.

7
PLANET PLUNDER

In 1968 William Anders, the astronaut, took a photo of our planet from the surface of the moon. It was an extraordinarily beautiful photo of mostly water, deep blue oceans and expansive white clouds. The moment was captured as a remarkable combination of moon landing technological expertise and inspiring cosmic beauty. He later made the perfect explanatory statement. "We set out to explore the Moon, and instead discovered the Earth."

Several years later in 1979, a British scientist, James Lovelock, published the book *Gaia: A New Look at Life on Earth*. In line with the photo of shimmering Earth, Lovelock proposed that Planet Earth was not a lifeless, inert globe with all of its resources available for *man's* use, as Western and Enlightenment philosophy traditionally believed. It was, instead, a living planet featuring a critical life system at its surface. This natural system, Lovelock explained, regulates the life of Earth in the following manner: "it was not the biosphere alone that did the regulating but the whole thing, life, the air, the oceans, and the rocks. The entire surface of the Earth including life is a self-regulating entity." Lovelock went on to reflect that as he had continued writing: "I began to see us all, as part of the community of living things that unconsciously keep the Earth a comfortable home, and that we humans have no special rights, only obligations to the community of Gaia."[1]

But now we have a crisis. Over the past 50 years, as a society, those unfulfilled obligations to the community of Gaia, despite the warnings of the few who understood, have begun to unravel this critical life system at the Earth's surface. And only now have we begun to understand the critically dangerous outcomes of that oblivious inattention. But most of us still do not truly comprehend the sequence of what happened, how we arrived at this planetary debacle, or what to do now.

DOI: 10.4324/9781003489771-8

The following pages attempt to fill this critical gap in our collective understanding. However, this fourth ancient wrong presents itself somewhat differently than the previous wrongs. First of all, it is difficult to understand how a subject so technical can be successfully addressed by social integrative power and potential culture shifts. There are three factors that begin to explain this possibility. To begin with, all of the other ancient wrongs featured high levels of *negative difference* with subsidiary levels of *indifference*. But here with planet plunder and crisis, it is exactly the opposite—*indifference* is the primary transgressor, mainly because stopping environmental plunder challenges our way-of-life and comfort. There are two basic reasons for this: we didn't know enough before, and now the problem seems too large.

As a result, even the people we should be listening to are giving us very different messages. Greta Thunberg, a celebrated environmental activist, is saying *panic*; while Rebecca Solnit, a noted essayist, is saying *hope*. Although they seem, at first glance, to be different, they actually support each other. As Solnit advises: "Hope is not a lottery ticket you can sit on the sofa and clutch, feeling lucky. It is an axe you break down doors with in an emergency."[2] This chapter illustrates the environmental crisis, its interrelatedness to the other ancient wrongs, and how social power can work. But it also gives you the information that mends hope and panic together—because, yes, it is an emergency and we need an axe.

A Crisis of Rising Temperatures

Recent statements by the Intergovernmental Panel on Climate Change (IPCC) leave no doubt: "scientific evidence for warming of the climate system is unequivocal." According to documentation released in early 2023, the last 15 years are the hottest on record since such records were established 141 years ago in 1882. This warming trend, the result of human activity primarily in the wealthier countries of the Northern Hemisphere, began with the Industrial Revolution and the first use of coal, oil, and later gas. Now, the worldwide burning of these fossil fuels to produce electricity, warm and cool our homes, power industries, and run our cars, trains, and planes has brought us to the edge of catastrophe. If left unchecked, the carbon emissions from the burning of fossil fuels will create planetary conditions that will make it unlivable for most people and other living species.[3]

This use of fossil fuels since 1950 has greatly increased the amount of carbon dioxide in the atmosphere. Scientists found that in the past 1,000 years carbon dioxide (CO_2) varied little, ranging between 260 and 280 parts per million. But beginning with the Industrial Age in 1750, energy demand increased and the burning of fossil fuels increased proportionally. At first, the CO_2 augmentation was gradual: an increase of 30 parts per million took 200 years (rising 10%), taking us to the 1950s. But within these next 56 years, we have seen an immense increase in CO_2, now 417 parts per million and rising. This means carbon dioxide (CO_2), a greenhouse gas warming Earth's atmosphere, has now risen more than 50% within these past years. CO_2 is not the only greenhouse gas, however, which traps heat into the lower layers of Earth's atmosphere, causing rising temperatures and humidity; there are others. But CO_2 concentrations are the highest contributor, accounting for approximately 70% of the Earth's rising temperatures.

In this early part of the 2020 decade, the average world temperature was about 1°C (1.9°F) warmer than pre-industrial levels. While one degree doesn't seem like much, it has produced enormous consequences, including forest fires, droughts, extreme weather such as hurricanes and tornadoes, as well as melting ice caps (some as large as the state of Florida), flooding of coastal cities, die-offs of coral reefs, and heat intensification. As a result, the number of large-scale migrations of vulnerable people is precipitously rising. The list goes on. These tragedies are nearly all directly attributable to the burning of oil, gas, and oil by humans for energy production.

As a result, we entered this 2020 decade with less than ten years to do the necessary work to diminish this catastrophic temperature rise. This means keeping the maximum temperature rise of Planet Earth, now at 1°C, to *below 1.5°C as we enter the 2030 decade.* To do so, we must substantially roll back—by no later than 2030—the causative carbon dioxide (CO_2) emissions to under 400 ppm (parts per million) if we want to stop the worst catastrophes, including tipping points that cannot be reversed. With that accomplished, it will be essential to keep working for several more decades until carbon emissions are finally stabilized at 350 ppm and Earth's temperature returns to its long-term average. However, the bad news is that emissions levels, as of early 2024, are rising faster than initially predicted.

It's important to be precise here. We are not only treading on the well-appreciated grounds of science but also on James Lovelock's newly defined sacred ground of Gaia where "life is a self-regulating entity." The minute numbers and precision which create this "life entity" underwrite a beginning understanding on our part. Look at the basics: the air that we breathe, allowing life, is made up of 78% nitrogen and 21% oxygen, totaling 99%. The other 1% is made up of a mix of gases of which carbon dioxide is only one part. This is where the measurement of parts per million (PPM) becomes so important. Changing carbon dioxide levels from 270 to 417 *parts per million,* as part of the group of gases which provide *only* 1% of the life-giving air that we breathe, is an infinitesimally small change. But this change in the 1% of gases composing the air is currently changing the atmosphere from life-giving to life-threatening. This designates what can only be viewed as sacred, therefore requiring our protection. Lovelock's contention that humans have no special rights, but only obligations to the community that comprises Gaia, now begins to make better sense.

Despite this expanding knowledge of the critical balance required for life as a self-regulating entity, global consumption of fossil fuel energy has increased exponentially in the wealthier countries. Today, more than 84% of this energy comes from the burning of oil, gas, and coal (33% from oil, 27% from coal, and 24% from gas), accounting for approximately 84% of total energy use. Green energy consumption covers the remainder of the world's energy consumption at 16%: hydropower 6%; nuclear 4%; wind 2%; and solar 1%. This second group are the energy producers which *do not* produce greenhouse gases and are the necessary solution to climate change. Wind and solar energy-producing plants have been growing rapidly in recent years, offering extremely promising futures for rapid expansion if they are well supported. Hydropower, on the other hand, has a limited future because most suitable dam sites are in short supply, and environmental damages in recent years have caused reductions in its relative importance. Nuclear power, while it does not produce greenhouse gases, is also declining in *importance* because of continued issues of waste disposal and vulnerability to meltdowns.

The relative share of these renewables, not including nuclear, has doubled since 1970 to 11%. That doesn't sound like much but it is an excellent starting base, particularly since the oil, gas, and coal share has fallen slightly during the same time period. However, the fact that the total amount of fossil fuels continues to rise is what truly matters; that is what determines the amount of carbon dioxide (CO_2) in the air and our vulnerability to catastrophe.

The potential for going *above 1.5°C* also doesn't sound like much either, but it can and will create a cascading series of future events that cannot be stopped if we do not decide to be vigilant. This small difference in temperature will melt the ice in the Arctic Ocean and a large part of the Antarctic at a faster rate. The rise in sea levels will be irreversible and this rise will displace about 79 million people (equivalent to almost a quarter of the US population), all because of rising carbon emissions. But ocean researchers are now reminding us that there are other dangers also.

Many of the largest cities across the world, and in the United States, are situated next to oceans. With a 3°C increase, likely to be reached by 2050 if the temperature in 2030 reaches 2°C, we could be reaching the beginning collapse of civilization. A billion people in places such as China, India, and Texas will find it impossible to work outdoors with summertime high temperatures exceeding 120°F—hotter than Death Valley. That is, unless we make some drastic changes.[4]

Carbon Corruption

They knew. The oil, gas, and coal industries all knew from the very beginning. A 1957 report from Humble Oil (now known as Exxon Mobil) tracked the enormous quantity of carbon dioxide" ... emitted "from the combustion of fossil fuels." The American Petroleum Institute's (API) *1968* report, conducted by the Stanford Research Institute, concluded that the burning of fossil fuels would bring "significant temperature changes by the year 2000." The oil industry may not have had access to the 1974 CIA Report which said the future economic and political costs of climate warming "would be almost beyond comprehension."[5] But if the oil industry did not have access, one has to ask: who wrote the report for the CIA?

So, why didn't the oil industry share this information? They were 'in the room' for every major policy discussion hosted by the Government and others throughout the 1960s, 1970s, and into the 1980s. But instead, while oil industry representatives were invited and attended every meeting or conference, the industry remained silent about their research. In trying to explain this situation, it brings to mind the just mentioned 1968 American Petroleum Institute report from one of the meetings their staff attended. At its conclusion, the Petroleum Institute authors noted what they termed an 'ironic' situation: "Politicians, regulators, and environmentalists were fixated on local levels of air pollution that were immediately observable, while the climate crisis, whose damage would be of far greater severity and scale, went entirely unheeded."[6] So, did they use this as their excuse?

But in the late 1980s, things changed for the fossil fuel industry. The 1988 testimony to Congress by the renowned NASA climate scientist, Dr. James Hansen, certified, without doubt, that greenhouse warming caused by carbon emissions was a true danger. In a Senate Hearing called by Senator Timothy Wirth, Hansen was the first scientist to go on record that the greenhouse gas effect was a *certainty*—global warming had begun. That message, with all of its specificity, made headlines across the country. After Hansen's report, the oil industry knew they could no longer sit on the sidelines. Instead, they went into high gear.

The oil industry immediately established the Global Climate Coalition (GCC) in 1989. The GCC mantra was that "the role of greenhouse gases in climate change is not well understood." But, as Green Peace explained, the GCC was created "to oppose mandatory reductions in carbon emission by obscuring the scientific understanding of fossil fuels impact on the climate." Two years later, The Union of Concerned Scientists reported that Exxon joined the infamous ALEC (American Legislative Exchange Council) "which actively undermines action on climate change at the federal and state level," and they worked together for the next two decades. Exxon only left ALEC under pressure in 2018.[7] Despite the dishonesty of the oil industry messaging, it did have a clear positive effect *for them* by raising uncertainty about the truth of the climate crisis.

In the 1990s, the oil industries escalated their now overt corruption. Rather than just continuing to sit on their damaging scientific reports, they began to actively lie about the planetary and human future. Part of the coal, oil, and gas industry's willingness to deceive was, and is, based on exorbitant profits. In most years, their profits had been between five and ten billion dollars annually.

But in the year 2000, that number went up to $17.7 billion.[8] The Obama Administration, recognizing the immensity of these profits, proposed in both 2010 and 2012 to eliminate 60% of all federal subsidies to coal, oil, and gas—$36.5 billion over ten years. But it was rejected—twice—by Congress, led by Senate Republicans. The Republicans claimed their vote was for the taxpayer, but it was a vote to keep oil profits up and, at the same time, keep the bottomless campaign contributions flowing to the Party. It continued: in 2021, Joe Manchin, a Democratic Senator from West Virginia, also operating with strong backing from the coal, gas, and oil industries, blocked the preferred Biden Administration environmental package from passing, in favor of a weakened one.

The oil industry has continued to leave nothing to chance. In 2015–2016 it spent $354 million in campaign contributions, of which 88% went to Republicans (I have not yet ascertained how much went to the Democratic Senator from West Virginia, Joe Manchin). So, it is no surprise in this 2020 decade that 97% of House Republicans now oppose taxing carbon pollution. Also, $354 million in campaign contributions is not a bad investment when, in return, the oil, gas, and coal consortiums are receiving a totality of *$29.4 billion* in federal subsidies, which comes to an approximate 8,200 percent annual return. And keep in mind these are *production* subsidies, not consumer subsidies. For consumer subsidies, the oil and gas industry receive another $14.5 billion (consumer subsidies lower the cost of fossil fuel use for its users). And just to be sure that the American taxpayer has paid the oil and gas industry sufficiently well, Congress continues its $2.5 billion annual subsidy for new fossil fuel exploration despite the fact that the IPCC is quite clear that no new fossil fuel resources should be opened if the 2030 climate target of 1.5°C is to be achieved.[9]

One more thing: these subsidies are more than *seven times larger* than the subsidies allocated to renewable energies all together. And let's not forget that there is no tax on "carbon pollution," a major degradation factor for our planet. The International Monetary Fund estimates this carbon pollution cost to be in the *trillions* of dollars when added in with subsidies.[10] It gets even worse if we compare it to other allocations of the federal government. Forbes magazine, a conservative-oriented business publication, headlined one of its major features in 2019: "United States Spends Ten Times More in Fossil Fuel Subsidies than Education." The article also adds that the fossil fuels subsidies are more than the country's annual defense budget which, as we all know, is no paragon of fiscal restraint.[11]

More recently, the oil, gas, and coal industry has ramped up its game to a new level of overt corruption. In January 2023, the Governor of Ohio signed a bill "to legally define "natural gas," actually a highly toxic methane gas, as a source of "green energy." This new legislation, supported by the gas industry, was coordinated by the infamous ALEC, made famous earlier by its links to the Koch brothers and the "dark money" political phenomenon.

Evidently, this "dark money" is at work again in a new category; we can now view their ads on TV urging all of us to adopt so-called 'natural energy.'

Even though promoted as "clean energy" or even "green energy," "natural gas" is not a renewable energy like solar, wind, and hydropower. It is, instead, a methane gas that actively contributes to greenhouse gases and carbon emissions. The so-called 'natural' gas is actually primarily methane gas (CH_4). When this methane gas is burned for energy, it produces about 117 pounds of CO_2 per million BTUs, as compared to 160 lbs. from liquid fuel oil and 200 lbs. from coal. So yes, this gas, supposedly called natural while actually methane, is slightly better than coal and oil, but it is still a major cause of greenhouse gases and carbon emissions. Further, methane gas is a by-product of not only oil drilling operations but also of farming operations that depend on cattle production. Together, they pose a critical but often unrecognized danger. Methane from all of the above production actions by humans has *80 times* the planet warming power of Carbon dioxide CO_2 in the near term, and requires equal attention.[12]

Climate activists are currently urging journalists and politicians to re-label, and call it "methane gas" since that is its primary component. But the Ohio Legislature defined the gas as green by stating in the bill that an "energy source can be considered green if it is more sustainable and reliable relative to some fossil fuels." Also, while natural gas does produce fewer emissions than burning coal, its production and use leads to multiple methane leaks. Even so, its backers added amendments to the Bill that made it easier to drill in state parks; and they admitted the bill was "stuffed like a chicken" when it hit the Governor's desk.

Of note, it was also known as the "chicken bill" as it was an amendment to a poultry bill. But more than that, it is well named because the oil, gas, and coal industry is playing chicken with our lives.[13] Should we remind these Ohio Republicans of our shared current reality? The Intergovernmental Panel on Climate Change in November 2022 clearly states: "to have hope of meeting the 1.5-degree warming goal, the world cannot build any more fossil fuel infrastructure."

Exxon-Mobil may be the poster child for fossil fuel promotion through cover-ups and deceptive means, but other US and foreign firms have followed along. In 2022 they all recorded record profits. Exxon reported profits of $56 billion and Chevron $37 billion, both US firms, while BP (British Petroleum) posted $28 billion and Shell (Dutch) $42 billion. These firms all previously announced plans to move *Beyond Petroleum* and invest part of their revenues in green energy, but they are now saying that stockholder votes have caused them to cut back on these plans. Chevron, for example, reported that in 2021 61% of their shareholders voted to embrace "green energy," but only 31% approved of that move in more recent polls. Clearly, the war in Ukraine has contributed to record profits, and changes in stockholder perspective have allowed the oil companies to take advantage.

An Exxon executive explained their company's latest position: "We leaned in when others leaned out," he said of the company's investments in boosting fossil fuel production.[14]

While some advocates argue the fossil fuel oil, gas, and coal industry is so big, and clearly has so much money, that they must be called upon to lead the transition to green energy. But that is like asking the frog to carry the scorpion across the pond. A better way forward would be to pressure banks, financial institutions, and pension funds to totally divest from petroleum companies and the industry as a whole. This means, of course, that a majority of stockholders would necessarily have to support investment in solar and wind technologies so that the oil industry could no longer use current stockholder sentiment, supposedly focused solely on profit, to hide behind and excuse themselves.

I think we all get the overall picture—subsidies should be abolished immediately and the supposedly natural methane gas should *never* be painted as renewable energy. Wind and solar power must become our first and immediate choice for a new energy infrastructure. Otherwise, we're just paying for the oil industry to ransack our planet and poison our communities. Of course, that is part of the culture shift—and changing thought-worlds—that are now necessary.

Delay and Denial

Many of us are latecomers to this difficult but necessary undertaking to repair the planet. It was only when the Paris Climate Meetings came on the scene in 2015, with 195 countries in attendance, that the reality of the environmental crisis was made real. Given this timing, it is surprising, shocking even, to learn that TV programs, discussion, and even classified reports about the climate heating phenomenon and its potential dangers actually began to appear as early as the late 1950s. There was, for example, a family film presented in 1958 by the Bell Science hour and the "bespectacled Dr. Research," the film's host who informs his audience about science's latest discovery:

> Man may be unwittingly changing the climate. A few degrees rise in the Earth's temperature would melt the polar ice caps … An island sea would fill a good portion of the Mississippi Valley. Tourists in glass-bottomed boats would be viewing the drowned towers of Miami through 150 feet of tropical water.[15]

That's pretty specific, isn't it? These types of revelations and information continued throughout the 1960s. Numerous books—starting with Rachel Carson's *Silent Spring* in 1962 focusing on the harm of pesticides, moving on to Barry Commoner's *Closing Circle* and Paul Ehrlich's *Population Bomb*—all began to strengthen environmental awareness. In 1969, the photos of planet Earth from the Moon landings, grabbed our attention *and* created a new and more conscious attachment to a fragile and beautiful planet. Then on April 22, 1970, Senator Gaylord Nelson established the first Earth Day by federal proclamation, intending to draw more attention to the "important issue of the Environment." And by the end of the year, President Nixon established the Environmental Protection Agency (EPA) "to protect human health and the environment."

During the following 1970s decade, five congressional Environmental Acts, critical to the clean-up of our then dirty and polluted environment, were enacted, four under the Nixon administration and one under Jimmy Carter. Today, they remain critical components of our present-day environmental institutions. These include: (i) the *Clean Water Act*, which regulates discharge of pollutants into US waters and defines quality standards for the country's surface waters; (ii) the *Resource Conservation and Recovery Act*, which governs the disposal of hazardous and non-hazardous solid waste; (iii) the *Toxic Substances Control Act*, which regulates the introduction of new or already existing chemicals; and (iv) the *Superfund Law*, to investigate and clean up sites contaminated with hazardous substances.[16] In 1977, President Jimmy Carter signed the Surface Mining and Reclamation Act to stop strip mining in Appalachia.

The 1980s continued to produce critical environmental research, international conferences, and numerous Congressional hearings. Yet, at the same time, according to many environmental experts assessing the situation currently, *little was done*. How can that be? Much was accomplished, but little was done. In fact, that is exactly the defining conundrum that has brought us to the environmental crisis we now face in this 2020 decade. Many useful actions were accomplished, but few were solidly focused on large-scale carbon emission reduction. And now we know why: the oil, gas, and coal industries did everything they could—including bribing the bribable politicians, as well as outright lying to us, the public.

As a result of this insufficient attention/action during the past 50 years, all of us have now entered the 2020 decade with less than ten years to substantially diminish this untenable and catastrophic temperature rise. But incomprehensibly, and dangerously, we seem to sometimes tolerate the same levels of delay and denial today in this 2020 decade as before. So, we should panic a little; but with the right actions, we can also learn to use that axe of hope.

History is an important part of this environmental narrative if we are to do more than just sit together and bemoan our collective fate or simply shrug our shoulders and pay no attention. Here, Nathaniel Rich in *Losing Earth* does us all a favor by using his historical lens to record these decades in detail.[17] In his analysis, I find three situations particularly detrimental and also formative of our current potential planetary debacle. At three critical points—in 1979, 1989, and 2009—we as a nation were prepared to act, and we quite shockingly did not. And in each instance, from an action perspective, it was the politicians, with pressure from the oil, gas, and coal industries, who backed off and did not act responsibly.

In **February 1979**, the first World Climate Conference, hosting attendees from 50 nations, announced a startling conclusion. The attending scientists came to a consensus and decided on a far-reaching solution: *a global treaty to curb carbon emissions*. At the same time, in 1979, one of the all-time climate crisis reports entitled "Carbon Dioxide and Climate: A Scientific Assessment" was published with broad scientific acceptance. The report was based on a collective review by the world's leading scientists and concluded that if no actions were taken, a three-degree warming of the Earth was catastrophically inescapable by the end of the 21st century. As a result, the US Congress signaled for the first time its willingness to act, and in 1979 President Jimmy Carter signed the Energy Security Act.

But later in that same year of 1979, Ronald Reagan was elected President, serving from 1980 to 1989. As a result, all the barriers to *any* environmental change immediately went up. Ronald Reagan's newly elected administration made very clear they wanted no part of any environmental activity because of its limitation on extraction for business profit from the existing environment. Reagan selected James Watt, president of a firm that fought for drilling and extraction rights on federal lands, to head the Interior Department, and Anne Gorsuch, a long-recognized anti-regulation "zealot" (and yes, mother of current Supreme Court Justice Gorsuch) was appointed head of EPA with the Administration's expectations that she would take the agency down from the inside. However, due to several unexpected situations, the 1979 World Climate Conference decision to hold a Global Summit in 1989, finalizing a binding international treaty to limit carbon emissions, remained on the table.[18] Everyone seemed to be on the same page, and everyone was expecting the US to lead in freezing emissions. But that did not happen as planned in 1989. Why?

In 1989 George H. Bush, newly installed as US president, succeeding Ronald Reagan, quickly discarded his own individual pre-election stance as an "environmentalist," and stood behind Reagan's initial hardcore opposition to the environment and regulation. He did so with strong support from the oil industry who welcomed him as one of their own. His appointment of John Sununu as Chief of Staff guaranteed a continuing anti-regulation and pro-growth stance, stifling the environmental movement at every turn. Sununu himself found James Hansen's scientific findings on global warming to be "technical garbage" and did his best to debunk them.[19] He was, in particular, determined to blunt the effect of the upcoming Netherlands 1989 Ministerial meeting. And he did.

On the last day of that international meeting, the final signing of the framework for a binding emissions treaty went late into the night. One American activist, gathered with others, awaiting the outcome at the entrance to the meeting chambers yelled to the Swedish Minister, "what's happening" as he passed by. "Your government," the minister said, "is fucking this up." And so, we did. Sununu's protégé had convinced Britain, Japan, and the Soviet Union to side with the United States and "abandon the commitment to freeze emissions."[20] During that year, 1989, at the Netherlands International Conference, carbon emissions were still at 353 parts per million. The nation and the world missed a critical chance, and critical time was wasted. As Nathaniel Rich reports, the 1989 consensus was strong: "Among scientists and world leaders, the sentiment was unanimous: Action had to be taken, and the United States was still expected to lead. It didn't."[21] Why?

In 2009 a third meeting took place during the administration of President Obama. Gone was the dream to freeze carbon emissions, however. Instead, this time it was organized around a less stringent necessity: to keep the Earth's temperature from rising to more than 2°C (3.6°F) above pre-industrial temperature levels. This time, it was the Obama administration that did not come through. They did not obstruct or bring down the agreement as the Bush administration had done in 1989. But they did *not* lead as they could have, and should have. Naomi Klein, in *This Changes Everything*, sees it as the "climate movement's coming of age: it was the moment when the realization truly sank in that no-one was coming to save us."[22] And so once again, to the great disappointment of the delegates and environmentalists around the world, even this level of maximum temperature was not made binding by the end of the 2009 Copenhagen Conference. Why?

The answer to each of those failures—the *why* of it all—is politics at its worst. In 1979, it was the newly installed Reagan administration's political preference, stating that "Government was the problem," which aligned perfectly with oil industry objectives. In 1989 it was again the same political preference by newly installed President George H. Bush, working with the growing hard-right of the Republican Party to continue government deregulation and market ascendancy.[23] Both of these preferences also aligned perfectly with the preference for executive privilege, discussed earlier in the chapter on Endless War.

But in 2009, it changed from political preference to political pressure. Within the Copenhagen Conference itself there were strong disagreements on how to move forward. In this context, President Obama's speech in the final hours of the Conference was a distinct disappointment to scientists and environmentalists. Although Obama said he believed the world was ready to "act boldly," he gave no indication that the United States itself was ready to "embrace bold measures."[24] Obama's clear preference was to act, but with the oil industry continuing to strong-arm the vacillating public, the President made a politically pressured decision to stand down.

But politics is not just about the politicians. In a democratic country, it is also about the people ourselves and how we vote. In the situations described above, it is clear that the Republican administration preferred deregulation, supporting the right of the oil industry to tell lies concerning the environmental damage they were causing. But let's be clear: it was the people who voted these administrations into power. So, either we, as the people, are willing to go along with the lies or we do not have enough initiative to seek out the facts. Even if some of us acknowledge the facts of the situation and vote to protect the planet, it remains difficult. There is among us a sufficient number of "Fence straddlers" who will align their vote with their own self-interest, creating and maintaining that 'moderate' political pressure. And politicians know that— even the good ones. So, in the end, the pressure to be "moderate," at least in some instances, also favors actions that protect the powerful and their profits.

Certainly, the oil, gas, and coal industries were playing dirty hardball then and continue to play dirty now; as are now the majority of Republicans. The moderate democrats, and in this case the Obama administration, both decided to play softball, acquiescing to political pressures when they should have demonstrated strong and solid leadership. When you do that, expectations are that you will do it again. So, later in 2012, when climate change was calculated to cost the US economy $240 billion a year and President Obama proposed lessening those federal tax subsidies given to the oil, gas, and coal industries, the Republicans and several 'moderate' Democrats once again said no. Delay and denial continue unabated.

New Global Frameworks

But there is good news and good progress. Despite the obstacles, the United Nations has moved forward to create a global institutional framework for climate change. In 1992. the "United Nations Framework on Climate Change" (UNFCC) was agreed to at the Rio de Janeiro meeting known as the "Earth Summit." It binds members "to act in the interests of human safety even in the face of scientific uncertainty." Today, it has near universal membership of 198 member countries. Eritrea became its latest member in February 2023. The three countries who are not members are Iran, Libya, and Yemen. The UNFCC Secretariat, located in Bonn, Germany, is responsible for ongoing management, including the annual meetings of the COP (members of UNFCC are referred to as "conference of the parties," or COP).

The UN-sponsored Paris Climate Meeting in 2015 was the success many had long been waiting for. With 195 countries in attendance, all agreed and signed the first universal, legally binding, global climate change agreement that begins to reduce carbon emissions. Known as the Paris Accords, it clearly states the basic objective: keep temperature rise below 2°C, and as close as possible to 1.5 degrees by 2030, while continuing to achieve the 2050 zero net carbon emissions goals. To assess this progress, each member country submits its internally agreed-upon goals and measurements to the IPCC annually, and the IPCC reports to the annual COP meetings.

It is useful to understand the scientific record here. Despite the multiple obstacles faced between 1989 and 2015, IPCC climate science diagnostics had been further sharpened. Data showed, for example, that global carbon emissions were *61* higher *percent* higher in 2013 than they were in 1990. That was the year after the infamous 1989 Copenhagen meeting blowout, originally slated to achieve this same goal, potentially making such a difference. So, in 2015, at the Paris meeting, reality finally began to sink in. It was finally recognized—and discussed at length—that if current emissions continued at that same pace, planetary temperatures would hit well beyond the previous expected benchmarks—somewhere between 3°C (5.4°F) and 4.4°C (8°F) by the end of this 21st century and well above 2°C by 2030. And this was one point that climate experts could all agree on—that level of temperature rise would be catastrophic![25] As a result, although the 2015 Paris Climate Conference began with the same 2°C (3.6°F) measurement as the 2009 meeting in Copenhagen, the situation soon changed.

Because of this situation, a group calling themselves 'The High Ambition Coalition' with John Kerry, the US representative to the 2015 Climate Conference, as one of its strongest leaders, began its advocacy. They pushed forward the adoption of no more than 1.5°C (2.7°) as the preferred maximum rise in planetary temperature by 2030. In the end, Kerry and the "High Ambition Coalition" did win. The 1.5°C temperature cap now serves as the Paris Agreement's preferred goal. As a result, this measurement stands as an effective organizing principle for the way forward.

Hope did take a momentary nosedive when it was reported in 2017 Carbon dioxide emissions increased after three years of stabilization. Luckily, and of extreme importance, the Intergovernmental Panel on Climate Change (IPCC) Panel Report verified, extrapolated, and explained the very real benefits for holding to the 1.5°C (2.7°F) goal in that 2018 Special Report, and this once again strengthened resolve of the national members. More good news was reported in the fall of 2020 when China announced its intention to achieve overall net zero emissions by 2060. Then, more good news: with the election of Joe Biden in November 2020, the United States quickly rejoined the Paris Climate Agreement, again with John Kerry at the helm.

But now there is a critical sticking point. What we are doing as a planet and as a country is still not enough, first according to the 2020 IPCC country revisions and end-of-year-2022 reports.[26] The bottom line at that time was we cannot afford more time lags. But in 2023, the reports from the COP made the bottom line even worse. They reported that while there was some progress in meeting poor country needs, the present pledges to reduce carbon in the atmosphere were insufficient, and global warming was progressing faster than anticipated. And in 2024 the prognosis of probable looming catastrophe only increased. This clearly means we should no longer allow oil, gas, and coal industries, or the conservative and moderate political forces that support their interests, to continue their environmental plunder and overtake the Gaia life-interests of the people and all planetary species.

In other words, if we don't stop this this clearly detrimental fourth act now, our children and grandchildren will look back and say—"why didn't they act?"

To avoid these extremely difficult climate tipping points, the goal is to arrive at 2030 with a temperature under 1.5°C. However, in January 2024, the European climate monitor Copernicus reported that the temperature hit 1.48 in late 2023. The Washington Post report said it well, but mildly. "The numbers might strike some as harbingers of doom. But they should not change people's understanding of the task at hand. Rather they should be heeded as a warning: If civilization wants to contain the damage from climate change, it must be willing to put more tools on the table."[27] But we must do more than that if we are to truly address environmental plunder in all of its multiple manifestations. Instead, all of us—but particularly those of us who live in wealthy countries—must do away with that self-serving acquiescence to extra-material comforts which does us no real favor, and allows indifference to Gaia's critical life system to spread and propagate.

The Inequalities of Inequities

Inequalities are now recognized as part of the potential climate debacle, one that we, as a high-emitting carbon emissions country, have let slide. We need to understand the specifics if we are to change the trajectory. While climate change is global in its impact, *energy use* and resulting CO_2 emissions are not evenly distributed among countries. The US ranks highest in emissions per capita of GDP (an average of 15 tons of CO_2 per capita, with an average income of $60K).[28] Other high emitters, countries such as Norway, Japan, and Sweden, emit less than 10 tons of CO_2 per capita and they also have slightly lower income per capita. China emits 7 tons per capita with a per capita income of only $16K. Also among the other high emitters are Saudi Arabia, Oman, and Kuwait, all with high incomes from oil sales but with little incentive to decrease fossil fuels, even with an abundance of solar power to potentially use as renewable energy.

It is also relevant to consider the per capita *consumption* of energy. Canada leads the world followed by the United States and Australia. China is in sixth place, behind Sweden, France, and Japan. China is the largest consumer of coal, while Sweden has opted for a relatively balanced mix of hydropower, gas, nuclear, and wind. Despite their relatively high consumption of energy, Sweden demonstrates that enlightened policies can be adopted to begin reducing greenhouse gases and their carbon emissions. Of interest, countries with a relatively high use of solar as a percentage of their total energy production include Japan with 5%, Australia with 5%, and Germany with 4%. Vietnam, however, leads the world at 6%, proving that even lower-income countries can achieve success with the right policies. Meanwhile, the United States use of solar is only 2%. Leading countries for wind include Sweden at 11%, Spain at 11%, Germany at 9%, and Greece at 9%. Wind makes up only 4% of the US share, the same as lower-income countries such as Turkey and China.[29]

There is no surprise here: the highest users of energy are countries with the highest GDPs/per capita incomes. This situation continues to exist despite the well-known fact that each country's national greenhouse gas emissions are too high to be sustainable. Meanwhile, the poorest half of the world have contributed very little to the excess greenhouse gases driving climate change. Americans, for example, emit more carbon dioxide in four days, on average, than people in poor countries such as Uganda or Ethiopia emit in an entire year. Also, it is too often the case that even within countries the difference between the wealthy and the poor is extreme. The richest 1% in the European Union, for example, emit 43 tons of CO_2 annually compared to a global average of five tons.[30]

But it is the poor countries and their people who are already suffering immeasurably from the effects of the global environmental crisis. *This suffering has nothing to do with their emissions,* only their geographical placement in hotter climate areas. For example, those majority populations across the world which depend upon rainfed agriculture, particularly in Africa and southeast Asia, are now living in drought conditions. Islands in the South Pacific or countries like Bangladesh are already losing land mass to rising seas; in 2018, 84 researchers studying West Antarctica proved conclusively "the frozen continent had lost three trillion tons of ice in the last three decades, further rising sea levels, with the rate *tripling* since 2012."

And we are only now beginning to include the outside costs of oil, coal, and gas impacts on ill health and danger. All of these conditions are poised to cause mass migration and further suffering on all sides. The suffering will be massive while posing highly increased national security conditions.[31]

But the poor country crisis continues unabated. In 2017 the UN made another mind-shaking announcement: after a decade of steady decline, malnourishment is once again on the rise, affecting up to 815 million people across the globe. Plus, there is the recent discovery that "rising carbon dioxide levels, by speeding plant growth, seems to have reduced the amount of protein in basic staple crops." In the hotter climes of the world where people rely on plants for their protein, that means huge a reduction in nutrition Add to this, the likelihood of being uprooted from one's home has increased by *sixty percent* compared with forty years ago

In addition, the burning of fossil fuels during extraction, or burning of fossil fuels in power plants, creates harmful air quality for neighboring communities (think 'sacrifice zones). In addition, the oil and gas used to cook and heat homes in every country is now finally recognized as increasing poor health (think inadequate fact publicity). Counted all together worldwide, the air pollution from fossil fuels kills 3.6 million people a year! This figure is six times higher than all the murders, war deaths, and terrorist attacks combined. Thus, to state that fossil fuels are a clear and present danger is a gross understatement.[32]

But it doesn't have to be this way. Overall, fossil fuels dominate the worldwide energy supply because they have been cheaper than all other sources of energy. But good news: in the past ten years, that has changed. The global average price of installed solar panels has fallen from $378/megawatt hour (MGh) of electricity to just $68 in 2019. Taking a levelized approach to compare energy costs, which considers not only the initial cost of erecting a power plant but also the lifetime costs of operations and maintenance, it is evident that solar and wind, essentially free, are now less costly than fossil fuel. Comparative costs worldwide average about $37 for solar (utility scale), $40 for wind, $59 for gas, $162 for coal, and $175 for nuclear/MWh.[33] The price of solar declined by 89% in ten years and is projected to go lower because of the economics of large-scale use. The cost of *off-shore* wind projects is projected to be even lower than those above. And finally, the cost of batteries has also seen a dramatic drop in price of 97% over the past three decades.[34]

Given this situation, it must be asked. Why is America not leading the world in carbon emission reduction? We are the biggest polluters, so that alone brings immense responsibility. But we also have the largest economy and an energetic and relatively well-trained workforce, so it means we can do more. Scaling up renewable energy is not only about lowering carbon emissions, however. It is also about cheaper energy and greater equity. Solar and wind power will lead to more jobs and cheaper prices for consumers at home. President Biden's 2022 "Inflation Reduction Act," which includes $369 billion to tackle climate change, places the United States close to achieving Biden's promise to halve CO_2 emissions levels by 2030, compared to 2005 levels. But we still have a way to go.[35]

There are new possibilities for building more stable and equitable economies by assisting poor countries to be explored. In the near future, the largest energy demand will not come from the currently rich countries but rather from the rapidly developing countries in Africa and Asia. The fact that most of these countries are in sunny climates is particularly conducive to solar power. Helping them take advantage of this resource is beneficial to all.

Low prices on green technology have made green energy (defined as wind and solar) the only and best solution for a viable future. This is equally true for both rich and poor countries of the world as a warming planetary atmosphere does not recognize country borders. So, these are all advantages to new alignments. Most important, we in the United States are also a democracy, so the knowledge of the *equality* necessary for equitable redistribution of energy resources is in our stated national ethos, if not yet sufficient in our practice. We need to act on that ethos. First, we can begin to practice equitable distribution of massive solar and wind power here in this country, thereby living up to our promised national and international obligations of strong and timely greenhouse gas reductions. We can then stand by our promised obligations in aiding the poorest countries deal with the already present damage of climate change. Other countries will follow.

Mending Hope and Panic Together

Obviously, there are lots of bad guys out there that we can blame for all of this. But the solution, as Naomi Klein said earlier, falls on us. It bears repeating: "It really is the case that we are on our own and any credible source of hope in this crisis will have to come from below."[36] And there are numerous environmental organizations, such as Sierra Club, that are emphasizing this focus. So, our first problem is to collectively decide *what* exactly is the problem. First of all, the conviction that the environmental crisis is out of our hands and beyond repair is grossly incorrect. Our shared environmental ethos about the nature of the problem has slowly changed. This emerging perspective has deepened as we defined the technicalities of the environmental crisis itself. Slowly, we are comprehending that while the climate crisis is necessarily scientifically defined, the *will* to solve it is people defined; and this people definition depends on *how we think about it*. This realization strengthens our emerging environmental ethos and adds to the present approach which remains almost entirely focused entirely on science, politics, and economics.

The key is turning around the *indifference* conserving the high level of superfluous comfort that we all have taken for granted. There were, and are, other reasons behind this indifference. "We did not know enough"; "the crisis was beyond our capacity to solve," and the concepts of uncertainty which were endlessly reiterated by fossil fuel advertising and bribable politicians. But these ideas and their supporting sub-conscious thought-worlds are now being set aside. Certainly, the environmental plunder leverage if we do nothing as Bill McKibben reminds us, remains with the big oil companies and those who associate with them. We can't let that happen.

Mending hope and panic together, more and more of us recognize that over the past 60 years, the unfulfilled obligations to the planet community of Gaia, despite the warnings of the few who understood, have steadily unraveled Planet Earth's critical life system. And we now comprehend the alarming outcomes of this dangerous inattention. But with that "axe of hope" in our hands, we now understand that social power and culture shifts will contribute critical work to solve this environmental crisis.

Two new alternatives to the environmental crisis have recently arrived, one in 2014 and one in 2018, which are in alignment with this shift in our environmental understanding. Both of these relatively new arrivals utilize emerging social and cultural formulations, now tentatively joining the existing scientific and economic realities. Together they introduce new ways to actually turn this environmental crisis around. The first arrival is the publication of a new book, *From Uneconomic Growth to a Steady-State Economy,* by the well-known ecology economist Herman Daly. It summarizes Daly's ground-breaking forty-year work on steady-state economics.[37] His basic premise that *economic systems are always a subset of the wider eco-systems* is essential to any positive environmental turn-around. In other words, Daly's organizing principle is the same one presented in *Gaia* by James Lovelock in 1979.

We have been reminded, however, of our responsibility for this "renewable quality of life" before; we just didn't listen. In 1977, during the decade of what could be called the aborted environmental awakening in the United States, the Iroquois Confederacy, also known as the Six Nations, sent a 21-person delegation to take part in the United Nations- associated Non-Governmental Organizations conference. Held in Geneva, Switzerland, the Iroquois delegation, as a nation, offered a series of position papers delineating the state of Planet Earth. Their presentation, entitled "A Basic Call to Consciousness: the Hau De No Sau Nee Address to the Western World," makes clear that human beings were not meant to abuse one another, or the planet. Pointing to the ongoing destruction of the natural world, they explain their liberation theology is a "*call to a basic consciousness which has ancient roots and ultra-modern, even futuristic manifestations.*"

> Liberation theologies are belief systems which challenge the assumption, widely held in the West, that the earth is simply a commodity which can be exploited thoughtlessly by humans for the purpose of material acquisition within an ever-expanding economic framework. A liberation theology will develop in people a consciousness that all life on is sacred and that the sacredness of life is the key to human freedom and survival .... It is the renewable quality of earth's ecosystems which makes life possible for human beings on this planet, and that if anything is sacred, if anything determines both quality and future possibility of life for our species on this planet, *it is that renewable quality of life.*[38]

This summarizes exactly what our environmental ethos must activate and become if we are to achieve that necessary culture shift to support "that renewable quality of life." Herman Daly's work as the founder of "ecological economics" fully embodies this principle, and his definitions of necessary actions provide a pathway to solving this environmental crisis. Daly, using economic analysis—sometimes dense—proves that it is necessary to recognize and live within this life-giving ecosystem. His latest book about steady state economics (SSE) clarifies the essential "renewable quality of life principle" for the regular reader and explains how it can be applied as a solution to the negative and brutal consequences of unchecked economic growth. The critical gift that Daly gives us is he redefines and realigns growth/economics within the deeper environmental ethos and culture we must develop.

Three points provide the essential foundation for this necessary change. Daly points out that immediate assumptions about steady-state economics tend to "block further thought." The first confusion is between *growth* and *development*. He states that a steady-state economy requires the cessation of growth ("accretion or assimilation of matter"), *not* of development ("qualitative improvement in design, technology, and ethical principles"). Daly goes on to observe that because current standard economics has become so "obsessed" with economic growth, it has failed to see that in some countries growth can be, and in some countries has become, *uneconomic*. Daly then states the second confusion to be avoided:

> [R]ecognize clearly that, in countries where poverty predominates, growth is still required along with development. To overcome poverty it is all the more necessary to move to a steady-state in wealthy countries in order to free up resources and ecological resources for the poor to grow into. Considerable sharing will be required to bring the poor to a level of sufficiency that is assumed for a steady-state economy.

Finally, Daly observes: "It is hard to imagine how a finite world of national economies all maximizing growth can manage to live in peace. Incentive to war is a major cost of growth that is not counted, and its reduction is an important argument in favor of a steady-state economy"[39] (think Putin's invasion of Ukraine). In other words, steady-state economics provides a foundation that is interactive with the world around it, rather than being impervious to its complexity demands. There are other books that build on the Daly complexity of it all. For example, *The Circular Economy: Challenges and Opportunities for Ethical and Sustainable Business* explains how the transition from a linear economy to a circular economy proposes to eliminate environmental damage.[40]

Does all of this sound too idealistic? Of course, we all know it's difficult to give up super-comfortable lives, and that's what it sounds like, doesn't it? But Rebecca Solnit once again gives us that forward-thinking perspective: "What if we were to prioritize reclaiming our time—to fret less about getting and spending—and instead 'spend' this precious resource on creative pursuits, on adventure and learning, on building stronger societies and being better citizens, on caring for the people (and other species and places) we love …?"[41] Doesn't that make living life with fewer 'things' seem adequately comfortable and, at the same time, so much more preferable? I like to think so; but how we 'think' makes a difference, doesn't it?

The second and collaborative alternative to solving the environmental crisis is found in the 2018 IPCC Special Report: *"Impacts of Global Warming of 1.5°C."* This Report is particularly important for a number of reasons, but primarily because it clarifies scientifically that the path forward to potentially limiting temperature rise to 1.5°C level by 2030 is indeed possible. In particular, in this Report the ability to remain below 1.5°C as we arrive at 2030 is also clearly linked to *social equality issues*. These links include "poverty eradication," "reducing inequalities," and "sustainable climate action" initiatives. In the Report each of these issues is assessed as positive, and is defined as an enabling condition to effectively limit global warming, particularly in this 2020 decade. It is important to note that these statements are not a preface to the scientific analysis; they are part of the scientific analysis.[42]

As a result, this IPCC Special Report, for the first time, aligns science with the new emerging environmental ethos. It does so by linking the social equality issues to the achievement of limiting temperature rise. As a result, it gives license in the scientific arena to use new and unexpected tools—social power and potential culture shifts—to finally stop the drastic unraveling of our planet's life systems.

The implications of this analysis, however, have not yet been fully recognized. But the fact that the IPCC 2018 Report's focus is clearly in accordance with Daly's Steady-State Economics, along with Lovelock's Gaia and the Six Nation Basic Call to Consciousness, should enhance this recognition of the role that equality plays and its implications. Because the IPCC scientists, in their reasoned scientific wisdom, have added their voices to this new equality alignment, a new environmental tool that is socially and culturally focused has now been defined. As a result, it can play a major role in diminishing temperature rise by 2030 to eradicate the unthinking and harmful practice of environmental inequality.

Together with Herman Daly's clarification that economic systems are always a subset of the wider eco-systems, this underscores the necessity for a steady-state economy which foregoes growth as primary. As a result, this expanding environmental ethos creates the tools for scientists to work hand in hand with all of us, the people, within a new social power framework, aligned with the existing scientific frameworks

But this is true only if we work fast. The most recent complete IPCC Report available at this point, released on March 20, 2023, combined the last six years of the UN Panel work. It reiterated and strengthened what had been already made public. IPCC scientists made clear once again how little time we have to make things right: "There is a rapidly closing window of opportunity to secure a liveable and sustainable future for all," the IPCC authors write.[43] Now we know, given the 2024 reports, that this window is closing even faster than anticipated. And as I completed final editing of this chapter on July 21, 2024, it was just named the hottest day ever experienced by the entire planet.

Finally, can you imagine the oil, gas, and coal industrialists or their supporting politicians ever initiating this work? No, that is extremely unlikely; even their provisional support may be too late in coming. So once again, it is up to us, the people, working in deep participation groups and our various organizational alliances. Part of the strength of this emerging environmental ethos by which we move forward together is its inter-relatedness to the other ancient wrongs, which explains its potential for maximum impact. So, it is always useful to keep in mind why this maximum impact is possible. The words of Thomas Paine, used during our country's first revolution, explains: "We think with other thoughts than those we formerly used."

Notes

1 James Lovelock, *Gaia: A New Look at Life on Earth* (UK: Oxford University Press, 2016), xv.
2 Rebecca Solnit, *Hope in the Dark: Untold Histories, Wild Possibilities* (Chicago: Haymarket, 2016).
3 The wealthiest among us have acknowledged this probability. See Douglas Rushkoff, "The super-rich 'preppers' are planning to save themselves from the apocalypse", *The Guardian*, Sept. 4, 2022, https://theguardian.com/news
4 Mark Lynas, *Our Final Warning: Six Degrees of Climate Emergency* (New York: Harper Collins, 2021).
5 Nathaniel Rich, "Losing Earth: The Decade We Almost Stopped Climate Change", *New York Times Magazine*, Aug. 1, 2018, 11–16.
6 Nathaniel Rich, "Losing Earth", 20.
7 Sourcewatch, https://sourcewatch.org and Union of Concerned Scientists https://ucsusa.org have multiple articles on this issue.
8 Greenpeace, "Exxon's Climate Denial History: A Timeline", no date, https://gp.org
9 Dana Nuccitelli, "America Spends over $20 Billion per Year on Fossil Fuel Subsidies". *The Guardian*, July 2018, https://theguardian.com
10 Brad Plumer, "IMF Says We Spend $5.3 trillion on Fossil Fuel Subsidies: How Is that Possible?" Vox, May 2015, https://vox.com
11 James Ellsmoor, "United States Spends Ten Times More on Fiscal Fuel Subsidies Than Education", *Forbes Magazine*, June 2019, https://forbes.com
12 *Our World in Data*, 2022, https://ourworldindata.org
13 Maxine Joselow, "Dark money groups led Ohio to redefine gas as 'green'", *Washington Post*, Jan. 18, 2023, https://washingtonpost.com
14 "As Profits boom BP slows goals for Climate", *Washington Post*, Feb. 5, 2023, https://washingtonpost.com
15 Nathaniel Rich, "Losing Earth", 66.
16 Sarah Fecht, "What the First Earth Day Achieved" *Columbia Earth Institute*, Apr. 21, 2020, https://earth.columbia.edu
17 Nathaniel Rich, "Losing Earth", *New York Times* Magazine, 2018.
18 Nathaniel Rich explains in "Losing Earth" the convolutions of the Reagan administration which were full force against any support to the environment in its first four years, but in Reagan's second administration how it softened its objections due to the growing popularity of environmental action.
19 Nathaniel Rich, "Losing Earth", 2018.
20 Nathaniel Rich, "Losing Earth".
21 Nathaniel Rich, "Losing Earth", 6.
22 Naomi Klein, *This Changes Everything: Capitalism vs the Climate* (New York: Simon & Schuster, 2014), 11–12.
23 Nathaniel Rich, "Losing Earth", 2018.
24 Suzanne Goldenberg and Allegra Stratton, "Barack Obama's Speech disappoints and fuels frustration at Copenhagen", *The Guardian*, Dec. 2019, https://theguardian.com
25 Climate Action Tracker, begun in 2009 after the Copenhagen Environmental Summit, CAT is an independent source giving daily and monthly up-to-date analysis and information on all aspects of the Earth's environmental crisis. It provides clear graphs and explanation based on IPPCC science results, and is a highly recommended resource. https://climateactiontracker.org
26 Climate Action Tracker: IPCC country revisions, ongoing, https://climateactiontracker.org
27 Editorial, "Scientists Knew", *Washington Post*, Jan. 19, 2024, https://washingtonpost.com

28 These numbers are adjusted for trade, i.e., energy not sourced overseas
29 Hannah Ritchie, Max Roser, Pablo Rodado, "Our World in Data", 2022 online https://ourworldindata.org
30 Lazard's Levelized Cost of Energy Analysis, (version 14.0) 2020, https://lazard.com
31 Brian La Shier & James Stanish, "National Security Impacts of Climate Change" *EESI (Environment & Energy* 2017–2019, https://www.eesi.org/papers.
John Podesta, "The climate crisis, migration, and refugees". *Brookings Institute*, July 2019. https://brookings.edu
Christian Parenti, *Tropic of Chaos: Climate Change and the New Geography of Violence* (New York: Nation Books, 2011).
32 J. Leslieveld, K. Klingmuller, A. Pozzer, et al, "Effects of fossil fuels and total anthropogenic removal on public health and climate", *Proceedings of the National Academy of Sciences*, National Institute of Health, 2019, https://nih.gov
33 Costs for energy vary around the world depending largely on the availability of sources within a country, and also whether government subsidies are provided.
34 Lazard's Levelized Cost of Energy Analysis (version 14.0), 2020, https://lazard.com
35 Trevor Higgins, Sally Hardin, "Five major benefits of the Inflation Act's Climate Investments", *American Progress*, Aug. 2022, https://americanprogress.org
36 Naomi Klein, *This Changes Everything* (New York: Simon and Schuster, 2014) 12.
37 Herman Daly, *From Uneconomic Growth to a Steady State Economy*, Advances in Ecological Economics Series, Editors: J. Van Den Bergh and M. Ruth, Edward Elgar (UK: Edward Elgar Publishing, 2014), https://e-elgar.com
38 Akwesasne Notes, ed. *Basic Call to Consciousness* (Summertown, Tenn: Book Publishing Company, 1991), 117.
39 Daly, *From Uneconomic Growth to a Steady-State Economy*, viii–xiii.
40 Helen Kopnina and Kim Poldner eds. *Circular Economy: Challenges and Opportunities for Ethical and Sustainable Business* (New York: Routledge, 2021).
41 Rebecca Solnit, "What if climate change meant not doom—but abundance?" *Washington Post*, Mar. 19, 2023, https://washingtonpost.com
42 IPCC, "Summary for Policy Makers: Global Warming of 1.5°C", *An IPCC Special Report on the Impacts of Global Warming of 1.5°C. 2018* (United Nations, 2018), 18–19, https://ipcc.ch
43 Evan Bush and Denise Chow, "Major climate report issues a dire outlook for human life on Earth", *NBC News, Mar. 20, 2023.*

Bibliography

Akwesasne Notes, ed. *Basic Call to Consciousness*, Summertown, Tenn: Book Publishing Co, 1991.
Bush, Evan and Denise Chow. "Major climate report issues a dire outlook for human life on Earth", *NBC News*, Mar. 20, 2023, https://nbcnews.com
Climate Action tracker, ongoing online service, https://climateactiontracker.org
Daly, Herman. *From Uneconomic Growth to a Steady-State Economy*. UK: Edward Elgar Publishing Company, 2014.
Elsmoor, James. "United States Spends Ten Times More on Fiscal Fuel Subsidies than on Education: How is that Possible? Forbes Magazine, June, 2019, https://forbes.com
Fecht, Sarah. "What the First Earth Day Achieved". Columbia Earth Institute, April, 2020, https://earth.columbia.edu
Goldenberg, Suzanne, and Allegra Stratton. "Barack Obama's speech disappoints and fuels frustration at Copenhagen". The Guardian, December 2019, https://theguardian.com

Rex Curry, "Exon's Climate Denial History: A Timeline". *Greenpeace*. no date, https://gp.org

Halpern, Evan, and Aaron Gregg, "As Profits Boom BP Slows Goals for Climate". *Washington Post*. Feb. 5, 2024. https://washingtonpost.org

Higgins, Trevor and Sally Hardin. "Five major benefits of the Inflation Act's Climate Investments". *Center for American Progress*, August, 2022. https://americanprogress.org

Intergovernmental Panel on Climate Change. "Summary for Policy Makers: Global Warming of 1.5°C", *An IPCC Special Report on the Impacts of Global Warming of 1.5°C*. United Nations, 2018. https://ipcc.ch

Joselow, Maxine. "Dark money groups led Ohio to redefine gas as green", *Washington Post*, Jan. 18, 2023, https://washingtonpost.com

Klein, Naomi. *This Changes Everything: Capitalism vs the Climate*. New York: Simon and Schuster, 2014.

Kopnina, Helen, and Kim Poldner, eds. *Circular Economy: Challenges and Opportunities for Ethical and Sustainable Business*. New York: Routledge, 2021.

Lazard. "Levelized Cost of Energy Analysis" (version 14.0), 2020

La Shier, Brian, and James Stanish. "National Security Impacts of Climate Change". *Environment and Energy Institute*, 2017–2019, https://www.eesi.org/papers

Leslieveld, J., K. Klingmuller, A. Pozzer, et al. "Effects of fossil fuels and total anthropogenic removal on public health and climate", *Proceedings of the National Academy of Sciences*. National Institute of Health, 2019, https://nih.gov

Lovelock, James. *Gaia: A New Look at Life on Earth*. UK: Oxford University Press, 2016 (1979).

Lynas, Mark. *Our Final Warning: Six Degrees of Climate Emergency*. New York: Harper Collins, 2021.

Nucitelli, Dana. "America Spends Over $20 Billion per year on Fossil Fuel Subsidies". *The Guardian*, July 2018.

Our World in Data. UK: University of Oxford. https://ourworlddata.org, 2022.

Parenti, Christian. *Tropic of Chaos: Climate Change and the New Geography of Violence*. New York: Nation Books, 2011.

Plumer, Brad. "IMF Says We Spend $5.3 trillion on Fossil Fuel Subsidies", Vox, May 2015, https://voxnews.com

Podesta, John. "The climate crisis, migration, and refugees", *Brookings Institute*, July 2019. https://brookings.edu

Rich, Nathaniel. "Losing Earth: The Decade We Almost Stopped Climate Change", *New York Times Magazine*, Aug. 1, 2018.

Rushkoff, Douglas. "The super-rich 'preppers' are planning to save themselves from the apocalypse", *The Guardian*, Sept. 4, 2022, https://theguardian.com

Solnit, Rebecca. *Hope in the Dark: Untold Histories, Wild Possibilities*. Chicago: Haymarket Books, 2016.

Solnit, Rebecca. "What if Climate Change Meant Not Doom?" *Washington Post*, Mar. 19, 2023, https://washingtonpost.org

Union of Concerned Scientists. *Climate Deception Dossiers*, 2015–2007. https://ucsusa.org

PART II
Activating Social Power

8
WHY SOCIAL POWER?

It is not easy to read about the four ancient wrongs. Their diverse manifestations in our current world are painful to fully comprehend. So, to stay positive and keep moving forward, we need to have some understanding of why and how these ancient wrongs, profound as they are, can be so effectively and completely dissolved. Actually, it's relatively easy, particularly in times of instability.

It is only social integrative power (also called social power) that can bestow legitimacy—the license to operate in society. No social order, comprised of system, structure, institutions, and organizations, can survive long without this socially bestowed legitimacy. Here, the essential key is that the deep participation methodology changes both the game and the odds. It does so quite simply: it necessarily moves the game from political/economic power to social integrative power, where legitimacy lies. When the interface between political/economic power and social integrative power gets out of sync, trust is degraded. Because it is only social power which can *create* the trust, which political/economic power depends upon, social power is the critical change agent. Thus, moving to social power increases the odds of achieving new legitimacies which neither political nor economic power can bestow.

The question of *how* social power creates and conserves *trust* is explored in Part III. But in this Part II, we first explore two questions. First, exactly *why* social integrative power—with its capacity to both take away and create legitimacy—can defeat the four ancient wrongs as we know them today? And then, given its importance, why so little is known about social power itself? But we also must ask one more critical question. Why has it taken so long for all of us to recognize the acceptance still accorded to those legitimacies which have caused so much harm?

The following Part II chapters are designed to assist in activating social power to defeat the four ancient wrongs, and avoid as well, making mistakes that diminish the power's capacity. To begin, this Chapter 8 explores the relationship between social integrative power and legitimacy, and why social theory on this subject has been too often ignored it. Chapter 9 explores the collective memory-blocks and individual practices which continue to deny social power activation. Chapter 10 reveals the economic and political institutional practices that keep the old legitimacies alive, and the through-line of inequality active. Chapter 11 identifies the collective negative thought-worlds which support this same inequality system and ascertains the two critical change dimensions by which we can collectively dismantle this long-enduring inequality system.

This specific chapter first introduces the results of Part I's investigation of the four ancient wrongs and how they continue to manifest in society. That is our starting point. Then we explore social power and why it works. But we begin with an acknowledgement that investigating the facts concerning Part I's four wrongs, as we have just completed, is an important first step. Redirecting our individual minds to begin questioning the collective worldviews of our society, allows us to recognize what we often thought was OK—what was considered to be "just and immutable"[1]—may not always be the actual reality. This Part II gives us insight into how these false realities have been maintained. More specifically, it gives insight into how the tools to undo those four ancient wrongs which still surround us can be identified and used effectively.

Investigation Results

We've made progress. Those earlier Part I initial questions have been helpfully answered: we know "where we are," and "how we got here." As a result, we begin to have some better understanding as to *why* these ancient wrongs, profound as they are in their current manifestations, can be so effectively overturned by social power. So, we are now better prepared to collectively move forward. However, moving forward does demand recognition of realities that we rarely think about which continue to block successful action.

Part I's investigation of the four ancient wrongs—the injustice of racism, class, gender; endless war; poverty; and environmental plunder—clearly identifies a *through-line of inequality* which connects together all of the four ancient wrongs as they now manifest. Intertwining each with the other, this through-line of inequality has been able to hide from sight primarily because of the multiple differing circumstances from which each wrong emerges. Because of this situation, *inequality* has escaped our best efforts to combat its inherent continuance. Equally important, because these inequalities run across the spectrum of our individual and collective affairs, we must now also question what was once unthinkable. Is our system's expressed values of freedom and equality, and their assumed underlying basic human morality, as strong as we have believed? Or is there a relationship between inequality and a right-to-command that has yet to be explored?

In this chapter, it was my original intention to catalogue the numerous instances of the inequality through-line running through each of the four wrongs and how it becomes interactive throughout. But I realized that it would be much better if you, the reader, or perhaps your deep participation group, working together, catalogue these yourselves. Otherwise, mutual learning and sharing doesn't work. In the coming Part II chapters, there will be multiple opportunities to read examples of how I see inequality running through and interacting. There will be, I expect, substantial agreements on the main themes. But at the same time, there will be differences in how they are viewed according to differences in perspective—age, culture, gender, locality—all of which are important to defining ways forward and, at the same time, better understanding what is happening now.

Currently, because of the failure to recognize the deeply embedded nature of inequality and the unconscious license given to it on our part as a society and community, we are now witnessing increasing numbers of more overt and obscene variations of inequality's violence play out in front of us every day. As a result, numerous people, even family, friends, and acquaintances, have contracted what seems to be inexplicable break-through infections of intolerance and even hate. And, as the manifestation of the four wrongs becomes more apparent, lies and deception have also come to dominate much of public life. All of this illustrates how the elements of this historical system—the structural, the institutional, and the organizational—assisted by *exclusionary powers* and *negative thought-worlds*, are still held captive to the practice of inequality despite a multitude of improvements.

Oddly enough, however, these inequality practices and even their obscenities are actually a sign of our success over the past decades. Despite the increasing violence and hatred which we are experiencing in this 2020 decade, at the same time numerous positive efforts have immensely expanded the philosophy and practice of *equality*, particularly and most overtly since the 1964 Civil Rights Act. Whether it is in terms of justice for all people, often pejoratively named the culture wars or identity politics; or protesting against illegal war; or working for labor rights; or raising awareness of our basic connection to planet earth; the continuing elaboration of these successes is everywhere and can only be characterized as extraordinary success.

And it is for this very reason that those leaders of *exclusion* and their followers—those who do not believe in equality in any shape or form—have decided to shed their sheep's clothing of supposed rationality and moderation. They now recognize the real *possibility of losing* their substantial political and economic power. They have realized that it is no longer effective to formulate equality solely as an aspiration for the future, or as a danger to the present, while expecting people to still believe in their exclusionary practices. Thus, tactics have changed. As a result, the exclusion leaders have become ever more willing to overtly use the underlying tools of domination and violence on anyone and everyone to maintain their hierarchies of supremacy. These tools, usually invisible to the established community but always used with impunity against those characterized as *less-than*, are now visible for everyone to see and fear.

So, odd as it may seem, this extremely difficult and even dangerous situation we find ourselves in indicates a true level of success. However, while this success deserves recognition, it is certainly not an enduring achievement. It is simply a way station which allows us to reflect and better understand our situation. Here, most important to consider is the concept of *equality* and its antithesis, the *practice of inequality*. Considering that the overt circumstances surrounding inequality practice for each of the four wrongs are decidedly different, it has been assumed, even with the introduction of intersectionality and critical race theory, that each injustice is separate from the other; in other words, more-or-less isolated each from the other. But it is useful to consider other possibilities. We begin to suspect that there may be a more profound reason for the inequality through-line's existence than the simple explanations of disparate individualized or group inequality acts. Instead, a shared subconscious belief system that fuels the inequality through-line across all four ancient wrongs becomes a possibility.

But wait a minute. You, as reader, may be thinking: "I'm not sure: that's taking a big jump." And you would be absolutely right in saying so. Moving from the possibility of individual actors coming together to maintain inequality and commit violence to the possibility of a covert, subconscious shared belief system that fuels inequality is definitely a big jump. Certainly, the different but simple circumstances supporting each of the four ancient wrongs are clearly identifiable and initially seem to support this individual actor reasoning. But it's also correct to ask—is that enough for its acceptance by all of us without further thought?

The answer is No; this observational point alone is not enough to come to such a conclusion. Instead, there are numerous research endeavors by multiple well-versed individuals on the theme of inequality, violence, and authoritarian rule which suggest a more interactive linkage, usually working itself through by subterfuge.[2] And in fact, when considered together rather than apart, the investigation of the four ancient wrongs strongly indicates such a probability. You may also remember reading in the Introduction to this book, how surprised I was that the violent situations I had observed while working in at-risk countries undergoing violence and rule-change would actually present themselves here in the United States. But these violent situations now exist here as well. So, both research and observation lead us to necessarily consider these more serious possibilities of shared subconscious beliefs.

Inter-Twining of Inequality

Given this more profound reason for inequality's staying power, there is a first question to ask ourselves: where would this subconscious belief and reliance on inequality come from, and why would it continue? A preliminary answer to this more complex question takes us back to our shared national history, where the damage was first done to our individual and collective humanity. It must be remembered that what we now call the United States did not begin as a democracy; instead, our country began as a *colony*. The casual extinction of the native peoples of this land; the suffering of the enslaved peoples from the African continent; the plight of waste people from England and the servant-slaves from its close-by Irish colony—none of this was an accident. It was colonization at work. The rules of the North American colony—like colonies everywhere—were brutally fashioned to consolidate profit and power by and for its ruling colonizer; in this case, the King of England and his men.

Most of us are seemingly disconnected from the reality of these earliest traumas of savage injustice generated by the concept of *negative difference and indifference*. Such a malevolent concept underlying every colony is difficult to grasp without experience. Recently, however, the 1619 Project, created by Nikole Hannah Jones and published in the New York Times, did bring the central history of slavery, underwritten by this concept of negative difference with all of its savagery, to the forefront of our national consciousness.[3] But our disconnect from these earliest traumas is not only expressed in terms of willingness to "forget to remember" this savage slavery along with its enduring racism but also our misunderstanding of the expansive breadth of negative difference and indifference.

To rule efficiently, development of *hierarchical exclusion* templates encoding the expansive, brutal, institutional preferences of the few in order to control the many was expressly created to conserve all types of power, wealth, and property for these few themselves. It is illustrated to this day, but at lesser levels, in every aspect of hierarchical domination belief and practice. This often-unrecognized type of domination is exemplified in each of the chapters of Part I, every chapter emphasizing different aspects of supremacy hierarchies with their necessary enforcement of belief in negative difference and indifference.

Achille Mbembe, renowned professor of history and philosophy, explains how this potent, but for the most part unremarked power of negative difference and indifference accompanying all aspects of inequality and hierarchy, actually works at an out-of-sight, sub-conscious level of any ascribing society or culture. Mbembe starts with the focus on enslavement of African peoples as representational of all brutal inequality practices around the world. He describes inequality's deep-dyed enduring strength by pointing to the fact that there is always a particular objective in mind whenever or wherever "negative difference and indifference" is installed.

Mbembe defines the objective:

The goal was to inscribe difference within a distinct institutional system in a way that forced it to operate within a fundamentally inegalitarian and hierarchical order.....

He then explains how this "distinct institutional system" works and endures:

Even though negative difference might be relativized ... it continues to "justify a relationship of inequality and the right to command."[4]

In other words, despite conscious changes initiated by cultural and societal demands against any particular wrong, negative "difference and indifference" can still maintain its intended inegalitarian system and hierarchical order at the underlying societal system level; to do so, it simply re-adjusts its pressure points when needed. Thus, separate fights against the four ancient wrongs here in the United States, or elsewhere for that matter, *don't, on their own, matter as much as we would like*. In fact, these individualized fights and movements actually account for the "progress and retrenchment cycle" that we continue to observe. Therefore, while equality may increase in one ancient wrong category or another, the essence and power of the inegalitarian and hierarchical order continue to remain intact. Also, because negative difference is operative and interconnected at the subconscious level but often unrecognized at the conscious level, inequality practice at the societal level continues. As Edward Said so saliently observed several decades ago in his book, *Culture and Imperialism,* social practice "acts as a cloak of respectability," allowing an injustice to remain conveniently hidden.[5]

Mbembe's explanation is an important one for two reasons. First, he critically defines what our investigation of the four ancient wrongs now identifies what is at work—undercover—here in our country. The continuing but separated practice of the four wrongs provides the fertile base required for the "relationship of inequality and the right to command" to maintain itself. As a result, economic and political power remain, hiding behind democracy, in the hands of a relatively small, self-proclaimed group of so-called heirs. Due to the ongoing actions of these so-called heirs, and despite our stated values of liberty *and* equality, inequality itself continues to be practiced freely. These inequality practices supporting the four ancient wrongs also shape-shift to counteract the changing demands initiated by those attempting to dismantle the four ancient wrongs in a one-by-one fashion.

Sadly, as a result, it's a bit like playing whack-a-mole for those of us attempting to expand equality. No matter *whom* or *what* is judged to be 'negatively different,' those judgments can be modified to 'acceptable' almost instantly when enough outside political pressure for greater justice is generated; then a level of success is felt and celebrated by those working for justice. But at the same time a new negative will be quickly invigorated internally within one of the other wrongs by that "distinct institutional system" described by Mbembe. Sadly, what has been so often missed by all of us playing this whack-a-mole game is that superficial changes do not change embedded exclusionary capacities.

The dawning realization of what this lack of honesty does to us as people and our overall societal system in terms of its structures, institutions, and organizations has been slow in coming, but it is here now. Allowing *inequality practices to flourish* without any real pushback or retribution creates a growing distrust of our national and local democratic institutions, as well as an attendant orientation toward corruption in our overall society. There are so many examples of this, but let's just take a few.

Part I's chapter on poverty indicated that both medical and poverty experts report that the United States is running a "high poverty economy" with little thought given to the "externalities" of hunger, trauma, homelessness, and lack of education. In this soul-eating situation, economic and political initiatives for change are intentionally limited by economic and political forces whose trustees have no wish to practice any sort of reciprocal care. And so, personal and public distrust, along with anomie, continue to grow like a fungus. And then, both the chapters on endless war and environmental plunder illustrate how institutions like the Defense Department, or gas and oil corporations, if allowed to operate with insufficient Congressional accountability harm those that they categorize as 'less-than,' but the practices themselves continue as acceptable practices. It's not just Congress that is recalcitrant in their duties, however. We, as citizens, too often gain from these corruptive practices or are too indifferent to demand some level of accountability. All of this allows distrust and corruption to grow and expand.

But we now have an option that moves us beyond working for the minimal change possible—and that is to actually confront those embedded negative thought-worlds and exclusions that keep these injustices and wrongdoing alive. This option is forged by local groups using deep participation to move themselves to the "revolution-of-thought" side of the ledger indicated several centuries ago by Thomas Paine. They make the critical difference by identifying those operative and sub-conscious negative thought-worlds still at work, and then working to remove the related inequality practices. It is here that social change reformulates itself as collaborative, social energy is introduced, and collective actions expand. By collectively embracing this option, we build on the change efforts begun long ago by those stalwart people and groups who continued to practice resistance with so much courage and so little ability to know if their efforts would truly achieve success. But because of their efforts, newly formed social action alliances are now able to re-organize themselves into a different kind of solidarity movement to not only resist but also reform, reinvent, and reimage. This reorganization begins the process necessary for culture shifts.

Current Social Thought

Social knowledge of society is required for students of sociology and social anthropology. It is also an imperative for those who want to change society for the better. So, for those who are just beginning to explore sociology and social thought, or those who wish to become more successful activists, or simply those who are observers and wish to better understand, a quick introduction to both the study of past social thought, and its roots for moving forward to the future, is essential. In these situations, I always refer to Robert Nisbet's *Sociology as an Art Form*. Nisbet gives a brief but overall synopsis of the nature of social reality. His first sentence questions the notion that sociology primarily is, or ever was primarily, a reductive discipline. He states: "[N]one of the great themes which have provided continuing challenge and also theoretical foundation for sociologists during the last century was ever reached through anything resembling what we are today fond of identifying as 'scientific method.'"[6]

Nisbet goes on to emphasize that sociology, with its focus on the illumination of social reality, is a form of study that moves hand-in-hand with art, both preceding it with discovery and following it with documentation promoting regeneration. He generally recognizes the diverse traditions of study within sociology and how they shape work, research, trends, and social action. The major traditions best described by Randall Collins are: the Conflict tradition; the Rational/Utilitarian tradition; the Micro/Interactionist tradition; and the Durkheimian tradition, also known as the Solidarity tradition.[7] The best recognized, and the dominant one is the *Conflict tradition*, but there are questions, raised in this chapter, about its sovereign applicability, particularly as societies move into instability.

Nisbet clearly relates sociology to capacity of the human mind to see, to reflect, and to connect. He looks beyond theory and empirical study to outline broad styles and themes. He captures sociological landscapes through image and symbol. He helpfully equates sociological portraits of the 19th and 20th centuries with "role-types." Finally, he captures the beginning factors of demise and transition in "the problem of motion" and "the rust of progress." Nisbet's description of sociology gives a useful glimpse of that stable world that is now disappearing, as well as the under-currents which have brought about this disappearance.

Finally, a critical concept—the foundation of Nisbet's book—explains how a book first written in 1976 can be so appropriate in explaining where the social sciences, particularly sociology and social anthropology, and even our everyday social world, find themselves in this 2020 decade. It is the connection he preliminarily identifies between sociology, the social, and "the nature of art and its relation to the human mind." Nisbet's pairing captures the necessity of *emotional resonance* and the *power of social relations* which exist between and among people, the planet, and all of its living beings. For example, one recent example of this capacity—in terms of art and its connection to the human mind and experience—are the paintings of Raqib Shaw. The artist's juxtaposition in the painting "Seeking Simurgh" depicts peace and an almost opulent tranquility inside, while looking outside at planes, bombs, and fire; revealing this 2020 decade with its divisions and its possibilities.[8]

Nisbet opened the door just slightly on these issues. But now, other social theorists continue to help us make that shift from 20th-century stability to the realities of 21st-century instability. Despite our historical system's ability to brush aside almost all dissent and criticism—either through outright dismissal or minimal reform—many social theorists and activists have persisted in pushing that door more fully open, as the bibliography of this book attests to.

It should be noted that those peering from the outside-in have made a particular contribution to our cumulative understanding of this shift. Achille Mbembe's *Critique of Black Reason*, V.Y. Mudimbe's *The Invention of Africa: Gnosis, Philosophy, and the Order of Knowledge*, and Edward Said's *Culture and Imperialism* all have had major impact. Each, in different ways, interrogates Western sociological images, illustrating how social and historical theory, working too often from its own Western internal orders of knowledge, can get it quite wrong, not only when investigating other societies but also when investigating their own society.

As a result, that motif of emotional resonance with its reality of social power lurking behind it, which Nisbet glimpsed decades ago, is now enhancing our understanding of current social themes, including authority, injustice, dislocation, de-colonization, the collective, and the sacred. Recognizing these affinities weaving their way throughout development of 20th- and early 21st century social theory allows consideration of how the study of society changes as it moves from stability to instability and then to demise and possibly a new genesis.

With this knowledge, we begin to recognize the problem. Neither the majority of social movements and protests concerned with desired change nor the preponderance of current sociological thought and theory have clearly recognized the implications of moving from a stable to an unstable society. There are, however, several long-term sociological perspectives and themes that zoom in on the diffuse problem of legitimacy and growing instability in different ways. In particular, the well-known American sociologist Immanuel

Wallerstein's definition and placement of social change within a historical system framework provides a starting point for these new instability and social power perspectives.[9]

Wallerstein first gives good news. In his book, *The End of the World as We Know It*, he discusses how the instability transitions now in process offer the best chance for change which can potentially reconcile "the search for truth and the search for goodness." Wallerstein also describes a critical base from which to proceed. He explains that every historical system, ours included, has three definable parts: (i) *genesis:* how it begins; (ii) *working system*: how it continues; and (iii) *demise*: how it ends. He maintains, and I believe many of us would agree, that we are now moving through the demise of our present historical system toward the unpredictable and unknown genesis of a new system.[10] He once summarized this process by observing that stable societies don't *budge*, while all that unstable societies need to change is a *nudge*.

Other sociologists, historical philosophers, and social theorists have broken through the assumption of stability to potentially explore moving through demise, but it's not necessarily positive or helpful. For example, post-structural inquiries, made famous by historian and philosopher Michel Foucault, with his substantive critiques of historical reason and societal truths, while insightful, has also successfully ushered in an age where all legitimacies are negatively questioned if not debased, with the role of the social itself often questioned. Whether Foucault intended so or not, high levels of cynicism about all societal institutions have now evolved.

As a result, the concept of society itself has lost some of its relevance and legitimacy in the past several decades. Poststructuralists, who in the sociology world are now described as *Rational/Utilitarians*, have attempted to further marginalize various concepts of "the social." While many sociologists and social anthropologists continue to come out strongly against this "retreat" of the social, they also recognize the complex roots of necessary change. Andre Iteanu, Directeur de Recherche at the Centre national de la Recherche Scientifique, explains this perspective.

> The movement that criticizes [the social] does not simply rest on scientific grounds but constitutes above all a moral injunction. In its view, the notion of society is not only 'objectively' wrong, but it reveals as well the unacceptable moral posture of those who use it. To dismantle it thus becomes a radical act against all forms of colonialism and hegemonies. This specific form of radicalism ... has grown into deconstructing any sort of sociality.

Iteanu goes on to observe that this tearing down of social relations, using made-up artificial moralities to do so, creates a "fragmenting movement that threatens to leave us with nothing that can stand outside the individual, that is, with no social relations at all."[11]

Except for a few theorists, however, most of us tend to resist understanding, collectively and individually, what is really going on in terms of inciting instability transitions. But looking around, we begin to recognize the ever-increasing number of *contradictions* which tell the same story over and over, just in different circumstances. Those moments when contradictory events surprise us—those moments which should never happen according to our understanding of the world—are the moments when we briefly recognize the end of our society's stability, its long-term duration, and the beginning of demise.

Our current set of contradictions, from the past 60 years or so, continues to multiply—a few enlightening, but many more inciting fear and instability. The 1960s are replete with them, starting with a devastating assassination of a president, followed by an unexpected uplifting with the Poor People's March on Washington. In the 1970s, we cannot forget the Vietnam War photo of the young on-fire Vietnamese girl running straight toward us after being hit by napalm. In the 1980s, for the first time in human history, the icecaps of Antarctica began to melt. In the 1990s, we witnessed the almost overnight dissolution of the Soviet empire, US's Desert Storm War against Iraq, and the first high school mass shooting at Columbine High School in Colorado. We will never forget the 9/11 2001 World Trade Center and Pentagon attack, with 2,977 dead. Beginning in 2012, we witnessed, dumbstruck, that first elementary school mass shooting in Newtown, and then its unrelenting continuance—Parkland High School, Uvalde Grade School …

But shockingly, the contradictions now come even faster. In 2020, we saw that knee on George Floyd's neck again and again, reminding us that we as a society had not been sufficiently aware and now can no longer even keep track of the number of young black men being killed. Overall, in 2023, there were more than 650 mass shooting in the United States.[12] And now, moving from 2023 into 2024, the obscenities of war and fear flare again, bringing destruction of life and cities with no end in sight. During this same time period, in the midst of the 2023 beautiful East Coast summer, communities were enveloped by smoke from a Canada wildfire almost 1,000 miles away, wiping away complacency that environmental crises would occur elsewhere and not be shared.

Finally, realization begins to set in. All of these events—with their increasing intensity and incoherence—are all connected, but we just haven't paid sufficient attention. Now, recognizing the through-line of inequality with its underlying reality of negative difference and indifference continuing to support hierarchy, we begin to glimpse these connections. The numerous contradictions we are now experiencing throughout our society in those blink-of-an-eye moments, as we register them, begin to unveil new realities and insights. We begin to understand that our attention is mandatory. We are finally catching up to reality.

From Demise to Genesis

The world-renowned theoretical physicist, David Bohm, with his surprising but very real interest in the social world, emphasizes two points of view which are critical to interpreting these surprising contradictions in a world that has moved from stability to instability. The first is the importance of what he calls *"participatory attention"* to explain our world. As discussed further in Part III, Bohm explains that to justly solve world problems, we must begin to share a common content even if we are not in entire agreement.[13] But first, Bohm's critical observations on the "interplay between our beliefs and what we experience as reality" delineate how shared beliefs and realities tend to go wobbly during demise transitions. Bohm explains the resulting reality this way.

> Reality is what we take to be true. What we take to be true is what we believe. What we believe is based upon our perceptions. What we perceive depends on what we look for. What we look for depends on what we think. What we think depends on what we perceive. What we perceive determines what we believe. What we believe determines what we take to be true. What we take to be true is our reality (emphasis added).[14]

People's wobbly truth and reality orientations begin to grow as instability picks up speed and moves through demise. This is because "what we look for" often changes as people's trust in, and the connections to, institutions of society diminish. As a result of this increasing mistrust, people and sometimes entire communities begin relying more on those subconscious negative *thought-worlds* which are always there to offer the comfort of stability in the face of change, even though they may be intrepidly wrong and unjust. As a result, trust continues to diminish, and divisions among people and groups expand.

Wallerstein also voices similar concerns to Bohm's about reality and truth, but from the vantage point of exploring the positive side of stability to instability, or demise-to-genesis possibilities. In doing so, he emphasizes what he sees as the responsibilities of social scientists. He first points out their fears of exploring these transitions. By doing so, he highlights the similarity between his and Bohm's perspectives. In Bohm's explanation, because of increasing mistrust, many people begin relying on their inherited perceptions. In this situation, those thought-worlds, particularly the negative ones, begin to have greater power to shape what is looked for.

Wallerstein describes something similar, but more specific. He notes an essentially unacknowledged reluctance on the part of social scientists to dig down into the roots of society's current perceptions of mistrust. Wallerstein infers this reluctance is because the social scientists themselves share, as we all do, high levels of unwillingness to leave the stability of the established legitimacies. But then Wallerstein turns to the positive. To explore "real historical alternatives," he suggests that social scientists should take the lead in actually sorting out why people actually distrust the existing political, economic, and social institutions. This is how Wallerstein explains it.

> "The possible is richer than the real." Who should know this better than social scientists? Why are we so afraid of discussing the possible, of analyzing the possible, of exploring the possible? We must move not utopias, but utopistics, to the center of social science. Utopistics is the analysis of possible utopias, their limitations, and the constraints on achieving them. It is the analytic study of real historical alternatives in the present. It is the reconciliation of the search for truth and the search for goodness.[15]

Like David Bohm and Immanuel Wallerstein, most of us want to believe that change can bring forth greater justice. But effectively exploring our present reality in relation to utopistics, with all of its implications as Wallerstein suggested, requires something more. It means a willingness to discover and acknowledge the failures of our current system which is difficult to do, primarily because it requires stepping outside society's acceptance circle. But any participatory and collective endeavor to describe a successful move from demise to new genesis landscapes demands exactly that.

Moving from the demise of a once stable societal system, through ensuing instability, to the genesis of a new social system is a challenging process, to say the least. So, it's important to back up for a minute and understand how a successful move such as this actually evolves. First, it's relevant to acknowledge that inside and outside of sociology circles, relatively little is known about social power, particularly in comparison with economic exchange power or political threat power. The enduring notion of a "stable society and its resilience" begins to explain why this is true and why sociologists and social change experts have often ignored it.

When societies are stable and collective sets of conscious beliefs and mindsets seem permanently established, social power retreats and takes a backseat, leading to the perception of its unimportance. In this situation, societal problems are at the forefront, while the underlying socially legitimated historical system of society with its structure and institutions, supported by various numbers and kinds of thought-worlds, is rarely recognized. But this subconscious, basic societal system, unrecognized as it may be, continues with its previous conscious legitimacies intact because its structures, guiding institutions, and negative thought-worlds have been insufficiently collectively identified, questioned or challenged. That is, until the moment when the contradictions become so numerous that they must finally be acknowledged—like now in this 2020 decade.

But until the very end moment, *instability* is easy to ignore. Social systems, once legitimated, rely on a kind of mutuality and inherent politeness which makes it difficult to step outside that designated circle of acceptance even when there are indications of violence and inequality practice. For all of these same reasons, it is extremely difficult to challenge and change existing legitimacies, even as we edge toward instability. This overriding inertia of stability explains, to a certain extent, sociology's focus on the *problems* of society with its cyclical changes. Addressing society's problems is easier to comprehend and take action than questioning society's actual organization and its subconscious adherence to long abiding *thought-worlds* that may be negative. For all of those reasons, challenging long-standing legitimacies underwriting society and its powerbase is undertaken by only a few in times of stability, but numbers increase as instability rises.

Given this situation, it is useful to refer to Kenneth Boulding, a well-known economist, for his perspectives on this subject. Boulding is perhaps best recognized as creator of the iconic phrase *spaceship earth*. In a 1966 article, he articulated, well before other economists, or conservationists for that matter, an understanding of the imminent climate crisis. Boulding has also been ahead of his time on other issues, including his interest in social integrative power. He explains this perspective in one of his last books, *Three Faces of Power*.

> My major thesis is that integrative power is the most dominant and significant form of power, in the sense that neither [political] threat power nor economic power can achieve very much in the absence of legitimacy, which is one of the more important aspects of integrative power. Without legitimacy both threat and riches are "naked." The great fallacy, especially of political thinking in regard to power, is to elevate threat power to the position of dominance, which it does not really possess. Failure to understand this is an enormous source of error in human decisions, both at the individual level and at the level of those who control organizations.[16]

Therefore, we begin to understand that legitimacy—even in the background—remains an essential factor in the organization and maintenance of the social order which surrounds our lives. When legitimacy and trust begin to be unexpectedly torn down by lies, exclusion, and violence, as is happening in this 2020 decade, realization comes slowly because there is an assumption of resilience and that repair will be relatively easy. That is not true; partial destruction often means complete destruction. To rebuild, reinvent, or reimage, to move from demise to a new genesis, the *reinvention of legitimacy* takes inclusion, critical thinking, collective social energy, and group solidarity to construct or reconstruct new ways of thinking. This social process is essential if the underlying changes to social systems and structures are to be recognized as legitimate, and consequently given the license to operate. This is the real power of the people: without social power, there is no legitimacy, no license. But social power is our power; it is the people's power!

The Power of the Social

For the most part, we all accept that cultures, community beliefs, political, and economic systems are socially constructed sets of societal institutions.[17] This has been further emphasized by the now widespread understanding of "institutions" as a concept, in large part due to the awards given during the 1990s to three Economic Nobel Prize winners—Douglass North, Amartya Sen, and Joseph Stiglitz. Although each of these Nobel Prize winners was working from a different perspective, the social and institutional context within which economics works was clarified for the first time by all three. This allowed terminology, long used by sociologists and anthropologists, to become widely recognized. Douglass North's definitions, in particular, "institutions are the rules of the game" and "organizations are the players," have become standard fare.

What is not so often clearly understood is that these now well recognized institutions also have an *underlying legitimating societal system*. Kenneth Boulding further explains the importance of social integrative power. He explains that social power operating at this underlying level produces critical legitimacies associated with "status, identity, charisma, love, loyalty, benevolence, community, identification, and so on" for the working institutions that populate every society. But the most compelling part of Boulding's explanation is 'why' this social integrative power is important. He tells us: "*no institution, pattern of behavior, or role structure can exist very long without it.*" In other words, every institution, in any particular society, must be granted this collective social "license to operate" if it is to endure.[18]

For our purposes, it also makes sense to further distinguish between levels of societal institutions. There is a fair amount of agreement that all societies consist of three levels: individuals, organizations, and institutions. *Organizations* populate our everyday world and are the collective groups within which people take a variety of actions. *Working institutions* of multiple types provide the "rules of the game" and represent the values around which a particular society consciously organizes itself. Recognized institutions in the United States include, among multiple others, public education, home ownership, the economic market, the arts, and the private sector. The effectiveness of institutions is fairly easy to assess; do their initiatives and actions match the societal values they are meant to serve?

But there are also entire sets of institutions and structures that work behind the scenes without our conscious awareness. Social anthropologist Mary Douglas explains that such sub-conscious guiding institutions "leads perception and trains it and produces a stock of knowledge." These behind-the-scenes institutions can also be compared to Durkheim's "collective representations" or Ludwig Fleck's "thought styles." These societal institutions and their structures, because they are hidden and rarely recognized, are often described as "incognito." However, in times of increasing instability, the fundamental principles of social organization embedded in these structures and their accompanying thought-worlds become more visible. As a result, their organizing legitimacies become more apparent and are thus better open to challenge.

There is a particularly significant point to be kept in mind when working with these legitimating entities. Collective mindsets and shared thought worlds have a particular form of invisibility and therefore, protection. The most important thing to understand is that they all operate best off-stage. Research in the cognitive sciences backs this up. Neuroscientist David Eagleman explains that "the conscious mind is *not* at the center of the action in the brain. Instead, it is far out on a distant edge hearing but whispers of the activity."[19] Thus, "out of sight, out of mind" successfully describes the situation.

In these out-of-sight situations, we, as people of a particular society, are rarely aware of our own societal rules and assumptions. We just assume they are the normal and right; therefore, they are the unremarked and unquestioned way of doing things. But there are two basic problems with this situation of invisibility: The first problem is, because shared beliefs and assumptions remain outside the everyday individual or collective consciousness, they do not demand—they *never demand our recognition*. The second problem is related to the first. Because of this lack of demand for recognition and reflection, the mutual beliefs and assumptions that we may share as thought-worlds *have not necessarily incorporated any aspect of truth, morality, or ethics*. In other words, some of those long-forgotten thought-worlds and their encased beliefs and assumptions, if brought to light, will not necessarily be worthy of legitimacy in a democratic society.[20]

Thus, however dangerous or immoral, out-of-sight thought-worlds remain exactly that until an unanticipated situation or incident brings into focus an unseen and unremarked social rule. When this occurs, the background coherency of an individual's thought and life is challenged, and the involved person immediately takes notice. In the so-called normalcy of stable times, these episodes are briefly noted by the individual but soon put aside and forgotten. During these situations, many of us also tend to dismiss the cognitive and emotional burden of 'thinking it through.' Having the dependable support of unrecognized and therefore unchallenged thought-worlds, this thinking-through does not always present itself as a necessity.

In unstable times, this out-of-sight, out-of-mind posture is, however, persistently challenged. This experience may be individual, but it is more often mutually shared among larger social groups. In the spring of 2020, for example, the brutal murder of George Floyd shredded the last fragment of capacity for white society's belief in the fairness of police protection. It ripped apart a negative cultural coherency among the white population about the 'bad' wrongdoer vs. the always 'good' police that had endured, and promulgated injustice for far too many centuries.[21] Multiple social protests thus began.

But insightful moments are difficult to sustain. This therefore brings into question what it actually takes for a group, and then a society, to overturn negative belief systems and create a culture shift. For example, will the horror of George Floyd's murder and others endure and re-emerge to sustain the critical change necessary? Or will it once again flip back to out-of-sight, out-of-mind? The well-known sociologist Randall Collins compares these insightful moments to the breaking of a pane of glass—neither the actuality of the glass pane itself nor the truth of the situation is recognized until that particular moment the glass breaks. It is at this moment that awareness begins to rise that something is terribly wrong—what just happened?—and the lines of difference and indifference begin to tell their story.[22]

In situations such as these—of instability, disruption, and expanding awareness of injustice—attention to that underlying legitimating societal base, and the thought-worlds that sustain it, is critical. A clear intentionality is necessary if we are to focus on both the depth and placement of these thought-worlds, particularly if we are to take away that socially determined "license to operate" for those that are negative. But it may be easier than we think.

In these circumstances, social power, residing in the subconscious underlying our everyday world, begins to play a greater role. This socially formulated sub-conscious features three main entities: guiding social institutions, structural components of institutions, and a legitimating base system. Each is surrounded and affected by various thought-worlds—positive and negative. This resulting social power, cumulative in nature and operating together with deep participation, normally creates trust and belonging. In turn, it is these factors of trust and belonging which support and make up the interface between conscious and sub-conscious societal life. But when this interface between the conscious political/economic uses of power and the largely unrecognized social power gets out of sync trust and belonging are degraded and instability ensues.[23]

Just becoming aware that social integrative power exists, how it operates, and what it can accomplish, begins to make this social power more visible and accessible. And once social power becomes more visible and accessible, we begin to comprehend how critically important it is to take into consideration these changes in power, and what it means in times of instability to the social, political, and economic change necessary for culture shifts.

Within this same social power sphere, working through deep participation existing positive social legitimacies can be retained *and* new social legitimacies can be created. It is only here, between political/economic power and social integrative power, where the coalitions of trust and belonging reside, that these legitimacies can be re-invigorated and re-invented. When agreed-upon societal change is achieved using those power tools of deep participation and social power powerful horizontal *interface lines* are created, once again forming a strong foundational development for peaceful social, political, and economic development.

These same small but critical differences operate in everyday life too. But because we tend to notice and focus on political and economic power at every level little or no attention is paid to the reality of social integrative power. Despite this politically and economically temporarily-forced anonymity, social power is always there, maintaining the cohesion of diverse groups, multiple communities, and democratically agreed-upon actions. If we recognized and took better notice of social power in stable times, our societies and individual lives would most likely operate at happier and more successful levels. But we only notice social power when it's gone. And even then, we often have a hard time naming what has actually disappeared.

Questions

In this 2020 decade, we are now a society that is transitioning through instability—from that well-known and well researched stable *societal working system,* moving through *demise* and forward to a *new genesis.* All of these current changes and transitions create questions that we are just beginning to answer. Etienne Balibar, a French philosopher, several decades ago took on an 'elephant-in-the-room' inquiring question about the *Conflict tradition* in sociology, the most popular tradition among sociologists, and decidedly popular among other disciplines as well. Balibar courageously raised doubt about the basic precept concerning this well-accepted tradition. He specifically questioned the well-known concept: "the basis of social transformation is the transformation of labour." This concept accepts and uses conflict and competition as a basis for any intended transformation.

After posing the question, Balibar wondered whether he and others were "victims of a gigantic illusion regarding the meaning of their own analyses." So, he asked himself:

> Instead of representing the capitalist division of labor to ourselves as what founds or institutes human societies as relatively stable "collectivities," should we not conceive this as what *destroys* them? (emphasis included)

When I read that question for the first time, I was so happily surprised. Conflict theory, sometimes known for political economists as the Rational/Utilitarian tradition, continues to be the starting point for many internationally oriented social scientists working in various countries. So, I said to myself: "Finally! Someone is taking this concept on and questioning it from a sociological perspective, and not just an ideological one." But that wasn't quite correct. Later, after discussing the question in some detail, Balibar decided that because the question was so highly abstract, "it was better to redeploy the theoretical tools at our disposal" rather then than attempting an answer.[24] Actually, that made sense at that point in time; the question was indeed too abstract for that moment because *instability* itself was still quite mild. So, seeing beyond the problems of society, to its roots, would have been very difficult for everyone.

But my interest in this question continued. It intensified several decades later when I first began working in the Democratic Republic of Congo, a country suffering from intense conflict. As always, the conflict model remained, and remains, a strong tradition guiding much of sociological thought worldwide, and it may well deserve to remain so in *stable* societies. But I continued to have questions about what works best in other societies, particularly those that are experiencing ever-greater levels of instability. My participatory and social analysis work with communities, and even nation-states caused me to place myself solidly in the Durkheimian/Solidarity tradition exactly because of those questions.

I recognize that the Conflict discipline offers a strong basis for reform in stable societies. In these situations, conflict is primarily contained within rules of competition taking place under the guard-rails of society's agreed-to legitimacies and truths. But I also understand, based on experience, that when the stakes are more profound, particularly in societies experiencing instability and fundamental change, the Conflict discipline is exactly the *wrong one* to adopt. In these situations, conflict easily moves beyond any means of control; truth can become either anyone's guess, or the servant of the most powerful and violent. And yes, sadly, this now sounds familiar in our own country.

On the other hand, my own, along with many others preferred methods of social and participatory research based on Solidarity traditions, also have their drawbacks, particularly when community and/or national *legitimacies* are being challenged and broken. The sticking point is the same in either case. It is that the making of collective unity required by every society, first described in the Solidarity tradition by Emile Durkheim more than a century ago, while questioned, hasn't really changed.

The Solidarity tradition still supposedly depends upon what has been described as a "pre-rational" sacralization of legitimacy which does not include any aspect of conscious and collective critical thought. With that 'pre-rational' definition still included, the possibility of effectively assessing true human reality, particularly in situations of instability, is clearly thrown out the window and there is nowhere to go. So, neither my preferred Solidarity tradition nor the Conflict tradition in their entirety fit the realities of our current real world. But things are changing through participatory practice. What is needed now within sociology and all aspects of social thought, as discussed further in Chapter 13, is a recognition and recombination of traditions which combine solidarity and legitimacy together. This reformulation of the Solidarity tradition articulates a participatory social theory for a fast-changing unstable world.

To illustrate, it often takes living in unstable societies that struggle to reside within their agreed "rules of law" to appreciate how important "social legitimacy" actually is. This is because in stable countries, stable cities, or even stable neighborhoods which experience little conflict, the concept of *legitimacy*, once established, quickly loses its social tag. It is simply perused in terms of the political and economic. But in countries, societies, and neighborhoods that suffer from conflict where the rule of law is badly battered or nonexistent, sometimes it is only *social* legitimacy which can prevail. Given our country's recent slide into continuing and escalating conflict, it is important to understand the *co-existence* required for social power's trust and legitimacy. Learning through historical analysis is easier than moving elsewhere.

Notes

1. Fox Piven, Frances and Richard A. Cloward, *Poor People's Movement: Why They Succeed and How They Fail* (New York: Vintage Books, 1979), 12.
2. Ben-Ghiatt, Ruth. *Strongmen: How They Rise, Why They Succeed, How They Fail* (London: Profile Books, 2021).
   Applebaum, Anne. *Twilight of Democracy: The Seductive Lure of Authoritarianism* (New York: Knopf, 2021).
3. Hannah-Jones, Nikole. *The 1619 Project* (New York: One World, 2021).
4. Mbembe, Achille. *Critique of Black Reason* (Durham: Duke University Press, 2017), 86.
5. Said, Edward W. *Culture and Imperialism* (New York: Vintage Books, 1994).
6. Nisbet, Robert. *Sociology as an Art Form*, (New York: Oxford University Press, 1977), 1.
7. Collins, Randall. *Four Sociological Traditions* (New York: Oxford University Press, 1994).
8. Sebastian Smee. "I was sure I would hate Raqib Shaw's work", *Washington Post*, Mar. 3, 2024, https://washingtonpost.com
9. I should note that Wallerstein's concepts of instability transitions, coupled with my own experiences of working in countries in the midst of various kinds of instability, has influenced my own theoretical perspectives.
10. Hopkins, Terence K. and Immanuel Wallerstein, *coords*. *The Age of Transition: Trajectory of the World System, 1945-2025* (London: Zed Press, 1996). For a much shorter synopsis, see Wallerstein, Immanuel. *The End of the World as We Know It* (Minneapolis: University of Minneapolis Press, 1999), 19–29.
11. Kapferer, Bruce, ed. *The Retreat of the Social: The Rise and Rise of Reductionism*. Andre Iteanu, "When Nothing Stands Outside the Self (New York: Berghahn Books, 2005), 104–105, and 112.
12. Gun Violence Archive, "GVA: 10 Year Review". https://www.gunviolencearchive.org
13. Bohm, David, *On Dialogue* (New York: Routledge Classics, 1994), 30.
14. Bohm, David. Berkeley Lecture, 1977, Maria Popova, "Trailblazing Physicist David Bohm and Buddhist Monk Ricard on How We Shape What We Call Reality", *The Marginalian,* Annual Special, Favorite Books, 2023, https://themarginalian.org
15. Wallerstein, Immanuel. *The End of the World as We Know It*, 217.
16. Boulding, Kenneth E. *Three Faces of Power* (New York: Sage Publications, 1990), 10.

17. Hacking, Ian. *The Social Construction of What?* (Cambridge: Harvard University Press, 1999).
18. Boulding, Kenneth E. *The Economy of Love and Fear* (Belmont, CA: Wadsworth Publishing, 1999), 5.
19. Eagleman, David. *Incognito: The Secret Lives of the Brain* (New York: Vintage Books, 2012), 173.
20. Douglas, Mary. *How Institutions Think*, (Syracuse: University Press, 1986).
21. Reading books such *Killers of the Flower Moon* by David Grann brings this home with that emotional wallop.
22. Collins, Randall. *Four Sociological Traditions* (New York: Oxford University Press, 1994).
23. Douglas, Mary, *How Institutions Think*, 1986.
24. Balibar, Etienne and Immanuel Wallerstein. *Race, Nation, Class: Ambiguous Identities* (UK: Verso, 2011), 7.

Bibliography

Applebaum, Anne. *Twilight of Democracy: The Seductive Lure of Authoritarianism.* New York: Knopf, 2021.
Balibar, Etienne, and Immanuel Wallerstein. *Race, Nation, Class: Ambiguous Identities*. UK: Verso, 2011.
Ben-Ghiatt, Ruth. *Strongmen: How they Rise, Why They Succeed*, UK: Profile Books, 2021.
Bohm, David, *On Dialogue*, New York: Routledge Classics, 1994.
Bohm, David. Berkeley Lecture, 1977. Maria Popova, "Trailblazing Physicist David Bohm and Buddhist Monk Ricard on How We Shape What We Call Reality", *The Marginalian*, Favorite Books, 2023.
Boulding, Kenneth E. *Three Faces of Power*, New York: Sage Publications, 1990.
Boulding, Kenneth E. *The Economy of Love and Fear*, Belmont CA: Wadsworth Publishing, 1999.
Collins, Randall. *Four Sociological Traditions*, New York: Oxford University Press, 1994.
Fox Piven, Frances and Richard A. Cloward, *Poor People's Movement: Why They Succeed and How They Fail*, New York: Vintage Books, 1979.
Gun Violence Archive, "GVA: 10 Year Review". https://www.gunviolencearchive.org, 2024
Hannah-Jones, Nikole. *The 1619 Project*, New York: One World, 2021.
Hopkins, Terence K. and Immanuel Wallerstein, *coords. The Age of Transition: Trajectory of the World System, 1945–2025*, London: Zed Press, 1996.
Iteanu, Andre. "When Nothing Stands Outside the Self", Kapferer, Bruce, ed. *The Retreat of the Social: The Rise and Rise of Reductionism*, New York: Berghahn Books, 2005.
Mbembe, Achille. *Critique of Black Reason*, Durham: Duke University Press, 2017.
Mudimbe, V.Y. *The Invention of Africa: Gnosis, Philosophy, and the Order of Knowledge*, Indiana University Press, 1988.
Nisbet, Robert. *Sociology as an Art Form*, Oxford University Press, 1977
Said, Edward W. *Culture and Imperialism*, New York: Vintage Books, 1994.
Wallerstein, Immanuel. *The End of the World as We Know It: Social Science for the 21st Century*, University of Minnesota Press, 1999.

9
WHAT WE SAY IS NOT WHAT WE DO

The first sentence of *Capital and Ideology*, Thomas Piketty's best-selling book, states the following: "Every human society must justify its inequalities: unless reasons for them are found, the whole political and social edifice stands in danger of collapse." Piketty goes on to observe that while every regime has its weakness, "to survive it must permanently redefine itself, often by way of violent conflict, but also by availing itself of shared experience and knowledge." Some in our country have decided to opt for violence, the first inequality justification option defined by Piketty. But most of us prefer his second option, to redefine ourselves in terms of equality—"availing ourselves of shared experience and knowledge." In the case of violent conflict, all that is needed is anger and fear. But in this second case, clarity will be a necessity for the way forward.[1]

The intent of these Part II chapters is to expand that necessary clarity. The primary finding of the four ancient wrongs investigation, discussed in the previous chapter, is that a through-line of intertwining inequality runs through these wrongs as they manifest in our society today. Recognizing this inter-active dynamic gives a new understanding of the situation and begins to develop the shared knowledge required to collectively create the solidarity necessary for the practice of equality.

This selected method—availing ourselves of shared experience and knowledge—as Thomas Piketty suggests, allows each of us to bring our ideas, perceptions, and feelings to the table. What I bring to the table is the knowledge and practical practice of social power and social change, as it has been initiated and practiced in numerous communities and regions. As a result of this experience, I clearly acknowledge my strong emphasis, based on experience, of bringing together shared critical thought, collective social energy, and emotional resonance for change. So then, I also ask, what do you, as reader, want to bring? I ask this question because I envision a mutual discussion, and I hope you do the same.

We now begin to understand. The historical endurance of brutality for those branded as 'less-than'; the continuing illogic of doubletalk activated by willing politicians and economic power-holders; the crisscrossing roles played by most of us of 'Excluder' or 'Fence-straddler'; all of these inequality compromises contribute to the ongoing maintenance of a trust-breaking disorder. The singular cry for freedom, used for justification, has no merit without the accompaniment of equality.

In these Part II chapters, I discuss social power and social change practice in terms of diverse themes—individual, community, historical, political, global, and local variations. I urge you, as reader, to think about and bring your ideas and experience to this discussion. In that way, we can move toward a shared complex meaning of reality, rather than simple individual opinions of everyday life. I know, you're sitting there, and I'm sitting here, with much time and space between us. But the reflection effort is well-spent, either in reading and understanding the implications of this book or perhaps beginning to think about how you and the friends that you can have gathered together can best contribute to these new solidarity alliances and movements.

Excluders, Includers, and Fence-Straddlers

To clarify how social integrative power (or social power, if you prefer) actually works from the individual perspective, existing classifications of American political and economic ideologies are not only insufficient but often misleading. Instead, three *social* classifications—defined by people's belief in equality—are described here and used throughout the book. They are: *Excluders; Includers;* and *Fence-straddlers.*

On the one side, we have the hardcore *Excluders,* and on the other, we have the hardcore *Includers.* The hardcore *Excluders* (roughly 15% of the population based on my experience)[2] are very clear—they don't want equality in any shape or form. While this Excluder belief in superiority is often based on racism and, in this country, white supremacy, it can also be lodged in traditional beliefs of supremacist hierarchy. Thus, Excluders may be misogynists, for example, asserting the patriarchal traditions of a dominating and hierarchical society, or simply your basic authoritarian—where "the many serve the few." But one factor they all share is that hard-core Excluders have a long-lived antipathy to any sort of equality and therefore have no interest in social integrative power.

The hardcore *Includers* (also roughly 15%) strongly *believe* in equality as a human virtue and a necessary value of democracy. Despite their strong beliefs, however, most Includers have not yet sufficiently defined the necessary actions and practices that make *equality* a reality in social, political, and economic terms. This is for two reasons. First, the existing definitions of equality are actually quite thin, both here in the United States and worldwide, as they are rarely supplemented by documentation of ongoing practice. Second, and equally important, equality is looked upon as a goal rather than a current necessity. As a result of what can best be called shallow understanding, Includers therefore sometimes naively accept inappropriate compromises with the Excluders.

But it's not just divided into two groups. There is also the very large group of *Fence-straddlers* (roughly 70%) who vacillate back and forth across that clear demarcation line dividing Excluders and Includers. They do so by avoiding the facts and reality of a given situation, hiding it from themselves and others. In other words, they clean up reality until synonymous with what they want to believe. This allows them to fool themselves and believe that they hold most or all of the American democratic ideals. The moderates of both political parties, for example, often suffer from this orientation. But in certain situations, including culture shifts, the Fence-straddlers do shift to one side or the other and will often stay there on a particular issue.

Fence-straddlers are able to make these back-and-forth self-serving moves by avoiding the facts of a particular situation—*double-talking* among themselves and making it up as they go. As a result, reality for Fence-straddlers becomes what they need it to be in order to consciously continue serving their individual best interests. But to double-talk effectively they unknowingly and sub-consciously hold fast to long-held social beliefs, particularly of negative difference and scarcity. Often, that well-known dispensation—"it's just human nature"—as discussed in Chapters 2 and 4 is used to substantiate their perspectives.

Some, particularly those who have been economically, socially, or physically mistreated, will be the early adopters and exhibitors of that *wobbly relationship between truth and reality* mentioned in the previous chapter. This allows those groups, and other individuals that quickly align with them for their own self-interest as time goes on, to create those multiple rationalities and numerous false equivalencies currently on display in our society. This then produces further distance between reality and belief. The growing illogic then causes confusion, increasing fear, and with it, the probability of truly bizarre beliefs and possible violent actions, now observed almost daily as we move through this 2020 decade.

But that is not all. This particular position, held by the Fence-straddlers, is particularly exploited by the Excluders. Excluders take advantage of the internalized confusion suffered by the Fence-straddlers themselves, as we sadly see every day in the political arena. Generations of Excluders, using their special brand of 'double-talk,' have had one basic intention since the break-up of the hierarchical, authoritarian, dominating world they prefer. This is to confuse any issue connected to equality and true democracy by keeping those ancient, deceitfully wrong negative thought-worlds and their legitimacies in play. And they do a good job.

Of course, because you're reading this book, you probably prefer to be defined as an *Includer* and prefer to work as one. A transition to the Includer category is, however, neither easy nor automatic—we are a culture which utilizes an opposition stance easily. And when we take this opposition stance, we automatically see ourselves as the "good-guy." But truth be told, we (particularly those of us who think of ourselves as white) have been schooled to ignore the attendant exclusion that exists around this 'good-guy bad-guy' stance and its purported efforts to do good.

Despite these difficulties we—all of us together—are now in the excellent position of learning how to initiate social power and its transformations. However, there is more to learn for all of us of every color and hue, before we can automatically consider ourselves 'Includers.' The next four sections of this chapter describe how, despite our good intentions, we still let ourselves be tripped up, again and again by the Excluders and their go-along to get-along Fence-straddlers. It starts with the cleanup that has been done on our country's history by those Excluder double-talking turn-about experts and the lack of questioning by the rest of us. These same turn-about experts, relying on their calculated double-talk, interpret both national and local matters in a manner that benefits the few and not the many—but still, too often we tend to believe them.

In everyday life, for example, we too often also fool ourselves into supporting violent hierarchies masked, instead, as people and organizations saving community and society. And sometimes, we as individual do the wrong thing with ostensibly good intent. While all of this may sound negative, it's not really. Understanding how the Excluders, and their willing sycophants use their masked but still subjugating power to divert and diminish people's power is not only useful, but essential, in preparing ourselves to successfully activate the reality of social integrative power. The next four sections of this chapter are designed to assist us in gaining the shared societal experience and knowledge that Thomas Piketty reminds us is so necessary for living equality. They each explore, from differing perspectives, why the Excluders and Fence-straddlers continue to be so successful, and how we, as wanna-be Includers moving to true Includers, can instead gain that necessary culture shift momentum.

The Paradox Syndrome

The loudly proclaimed US motto "liberty and equality for all," and then in sotto voice – 'and prosperity too,' continues to be a mainstay of American culture. But this definition of the American dream also defines our national *paradox*. Along with this motto the accompanying wrongdoing charges of racism, sexism, classism, and war-mongering have also been repeated loud and clear over the decades and centuries. Oddly enough, these two labels seem to accompany each other. And that is why, together, the positive beliefs of liberty and equality, and the negative charges of wrongdoing add up to what can be best described as the *paradox syndrome*.

This phrase—paradox syndrome—with all of its subtle meanings and multiple interpretations begins to capture a reality that has allowed us, as a country, to not face the reality of that inequality through-line which permeates our society. Normally, people tend to attribute paradox practice as a form of individual *hypocrisy*; and it can certainly be that when practiced at a personal level. But when practiced at a country level—at a national level—it is something else entirely. It is no longer a personal hypocrisy, but instead, indicates a viable and overall *structure* of society.[3] This particular structural form is described as paradoxical because it presents a seemingly absurd statement that, when investigated, is found to be true.

The critical point here is that investigation is required for a paradox's truth to be uncovered and validated. But because the nature of paradox is constructed so that the true reality is difficult, if not almost impossible to dig out and truly see, we as Americans have lived comfortably within this syndrome, without questioning, for far too long. Now, however, we begin to understand: it is within this context that *what we say is not what we do* plays out its heart-wrenching resonance.

Political historian Heather Cox Richardson, in her 2020 book *How the South Won the Civil War*, identifies an encompassing reality of this paradox few of us have thought about. Bringing together the democracy principle and the little understood political paradox of belief in *both* equality and inequality, she identifies the first aspect of this paradox syndrome.

> America began with a great paradox. The same men who came up with the radical idea of constructing a nation of equality also owned slaves, thought Indians were savages, and considered women inferior. This apparent contradiction was not a flaw, though; it was a key feature of the new democratic republic. For the Founders, the concept that "all men are created equal" depended on the idea that the ringing phrase "all men" did not actually include everyone …. In the Founders' minds then, *the principle of equality depended upon inequality*.[4]

But there is more to it. Cox Richardson also identifies the paradox's corollary, in effect the consequence of belief in this principle—*equality will destroy liberty*.[5] When I read that phrase for the first time, I involuntarily sat back in my chair—I finally understood. This principle made sense in an intuitive, even emotional way, that all of the intellectual description did not… could not. I realized this is why so many people—ostensibly good people—grasp and hold ideas that are clearly unhinged from the truth, often hateful as well; and why some of them can be enticed so easily to commit violence.

Now it becomes clear: some people believe themselves to be injured because they presume their identity as a member-in-good-standing of the United States is being taken from them, and that means their freedom, their status, and their belonging is also being ripped away. This is what happens when the supposedly 'less-than'—people of color, multi-racial women, LGBTQ, and immigrants—are specifically identified and held as equal. Immediately, everything shifts into a negative space: this expanded designation of equality is exactly opposite to their sub-conscious thought-worlds (and sometimes, but not always, their conscious values) which allows them belief in their own particular and validated higher status. "Equality will destroy liberty" therefore symbolizes the introduction of inequality and a new 'less-than' identity. As a result, these injured groups of people are willing to protect their liberty and freedom, meaning their elevated identity and sense-of-belonging, at all cost against equality.

It is those Excluder turn-about experts, so adept at double-talking and managing the social psychology of meanness, who create the *delusion of injury*, and this delusion fuels the increasing violence around us. Mark Danner, book reviewer, describes it this way—"delusions have become our daily bread."[6] Of course, we can say, 'that's them, not us.' But that's not really true either; none of us are totally immune to feelings of injury. To completely cure this paradox syndrome we may need to recognize how we may have unwittingly participated in it. However, most of all, once we understand the genesis of these injured personas, we could say "no," that is not true at all; we are all equal here'. To say that with any truthfulness, we will also have to learn how to redesign our society for the equality that we have claimed for it for so long.

But hold on. There is also another aspect of the paradox to deal with, and it is much better known and sometimes even celebrated. According to Immanuel Wallerstein, the sociologist and economic historian we met in the last chapter, the economic aspect of the syndrome is one of the more basic *contradictions* of our current historical period. He perfectly zeroes in on the situation.

> We start with a seeming paradox. The major challenge to racism and sexism has been universalist beliefs; and the major challenge to universalism [concern for others and inclusion] has been to racist and sexist beliefs. We assume that the proponents of each set of beliefs are persons in opposite camps. Only occasionally do we allow ourselves that the enemy, as Pogo put it, is us; that most of us … find it perfectly possible to pursue both doctrines simultaneously.[7]

Wallerstein goes on to tell us that, similar to the political side of the paradox identified earlier, this economic side of the paradox is also not something to be attributed to the simple human failure of hypocrisy. Instead, he maintains that it is "enduring, widespread, and structural." The primary, overt job of this paradox, as Wallerstein explains it, is to keep the varying incognito economic *structures* built on inequality practice, and the well-defined conscious universalist *values* based on equality beliefs, well-apart and therefore hidden from each other.

Wallerstein doesn't buy the popular explanation that equality and universalism are winning the day either. Instead, he points to capitalism's "commodification of everything" as both the culprit and most likely winner. He explains that accomplishing the commodification goal requires a smooth 'flow' of commodified goods, and the existence of racism, sexism, and classism, all clearly demarcating "negative difference," usefully enhances the flow. This commodification does so by excusing *less than equal treatment* to those workers who are key to the production process and can be branded as 'different.'

Wallerstein explains that the market world legitimizes this situation through its philosophy of "emancipatory freedom and liberty" as necessary for the basis of exchange. Thus, at the local level, employers might say: "If people want to work, I am giving them something they want and need. If they want to take a low-paying, even dangerous job that's fine—it's up to them—they are free to do so. It's their choice, their responsibility. It's certainly no responsibility of mine."

Now, as we investigate these two seemingly absurd statements, the first which politically claims *"the principle of equality depends upon inequality,"* and the second which claims that it is perfectly possible for a democracy to pursue *less than equal treatment* for those labeled unequal and less-than, the true reality emerges. Both of these statements seem to be, at first glance, completely absurd. But it turns out, when investigated, that we, as a nation and people, have been living in a society whose societal structures formulate, shape, and maintain exactly these seeming absurdities—and we are only now beginning to see and recognize it. But we need to go further and recognize those people and groups that are continuing to present these dangerous absurdities as truth.

Turn-About Experts

Excluders with strong turn-about expertise are heavily represented at both the national and local levels. At the national level, they are particularly well-represented as lobbyists to Congress for those groups that are primarily interested in expanding their own power and wealth and typify to some extent both the economic and political sides of the paradox syndrome. The activities of these turn-about experts and the result of their double-talk for well-known industries—including armaments, oil and gas, big pharma, and social media/

AI—are well documented. These industries, in particular, are extraordinarily good at shaping the relevant national issues in a manner that benefits their interests and clearly hurts all of the rest of us. There are so many examples of turn-about success—for example, the immunity from prosecution which gun manufacturers and social media still enjoy in most states. Another example of this turn-about expertise are the huge subsidies, mentioned earlier, that Congress continues to give to the oil and gas industry.

For all of those reasons, it makes sense to use one of these turn-about experts as an example of that capacity which continues to trip most of us up with some success. Every year, for example, as discussed in the Endless War chapter, Congress votes on the National Defense Authorization Act (NDAA), and preferably with little citizen attention, it passes with a strong bipartisan majority. As an example, for the 2022 NDAA the vote was 88-10 in the Senate and 363-70 in the House (51 were Democrats). The Bill itself totaled $777.7 *billion* for the Department of Defense (DOD) and ancillary departments, including the Department of Energy for nuclear oversight. To be clear, this did not include the *cost of war*, only its equipment and support.[8] However, despite the fact that this budget allocation is just one item, it is, at the same time 61% of our total national discretionary budget![i]

Another way to understand the major impact of this military and Department of Defense (DOD) spending budget on us as citizens is to review the corresponding *discretionary* side of the Federal budget, of which the DOD budget is a part. In comparison to the 61% of the 2019 discretionary budget allocated to the Department of Defense, the totality of all US government departments divided the remaining 39% in the following proportions:

> Government operations, 5%; Education, 5%; Diplomacy and foreign aid, 2%; Science, 2%; Veteran's benefits, 7% (although calculated separately, this is an indirect military expenditure); Energy & Environment, 2%; Unemployment and Labor, 2%; Health, 5%; and Housing & Community, 5%.

Is this an appropriate division of national resources? I don't think so, do you? But as a result of very effective double-talk by turn-about experts, aimed first at Congressional Representatives and secondarily at the rest of us, a massive amount of funding is allocated to DOD. In addition, this mammoth amount of funding is given with no demand for accountability, and assisted by the double-talk, which keeps the economic/political paradox syndrome alive and well.

---

i I use the 2019 budget as it is most in line with previous budgets. The 2020-2024 military budgets have been increased substantially, most dramatically since early 2022 through support to the Ukrainian War, and now to Israel for the Israeli/Hamas War.

So, why does this turn-about double-talk continue to work? Remember, in Chapter 5, Endless War, we discussed the fact that in 1990 Congress passed the "Chief Financial Officers Act" which requires all federal agencies to develop audit systems and subsequently submit to official government annual audits. All government agencies have been in compliance with this rule for years, except DOD. Finally, in 2018 Congress tried one more time. They hired a reputable audit firm, but once again, the Defense Department refused to comply, maintaining that the DOD organization was just too big and complex to complete the audit.

However, it is exactly this annual discretionary budget which we, through our elected representatives to Congress, supposedly control. The minimal, even miserly, allotments to the multiple government agencies not affiliated with the DOD illustrate how current military spending distorts justice and civilian well-being across the country. Facts to consider include the following. In 2016, the last year of the Obama administration, the National Defense Authorization Act was $618 *billion*. The 2022 NDAA, six years later, was at $777.7 *billion*, amounting to an increase of $26.5 *billion* per year, for a total of almost $160 *billion* since 2016.

That's not all; it's important to understand that this is not a political standoff. Nor does it have much to do with relatively short-term funding efforts, including the present short-term support to Ukraine. Instead, it's a long-term support policy on both sides, and they all play the same game. Moderate Democrats, for example, aligned themselves with Republican members of the Senate Armed Forces Committee in the summer of 2021 to *increase* President Biden's original 2022 NDAA request of $770 *billion*, adding another $25 *billion*. The result of this budget overkill was the funding requested by Senate members (Republicans and Democrats) resulted in 22 major additional war items (jets, ships, and super-hornets) being placed in the final NDAA, although not requested by the DOD. Senator Elizabeth Warren was the only Senate Armed Forces Committee member to vote against this $25 *billion* increase.[9]

How does this happen? How does—and let me use a proper academic word for this situation—absolutely *zilch* questioning and action take place concerning Defense Department spending? Yes, some senators and congressional representatives do vote against the bill, but they are a definitive minority; and yes, they do receive a certain number of emails and phone calls. But there is little or no substantive accountability on *any* side—Congress, citizens, or TV news media.

But wait—now we begin to have answers. We begin to recognize that economic self-interest is a paramount issue, hidden by patriotic doubletalk. But it goes much deeper than that. All of this is based on that two-sided political and economic paradox syndrome discussed at the beginning of this chapter. The *intersection* between those two—the twisting of political equality and inequality, intersecting with the economic claim that it's OK, even almost generous to hire a person in dire need for a dangerous job and pay them next to nothing. It is this intersection and confluence of energy between these two aspects of the paradox syndrome, with all of their surrounding negative thought-worlds inscribing that, as Achille Mbembe described earlier: inscribes "difference within a distinct institutional system," forcing it to "operate within a fundamentally inegalitarian and hierarchical order."

This energy trade between these two aspects of our national syndrome illustrates exactly how 'that unremarked power of negative difference and indifference, accompanying all aspects of inequality and hierarchy, actually works at an out-of-sight, sub-conscious level of any ascribing society or culture.' And as Achille Mbembe explains in the previous chapter, this out-of-sight "distinct institutional system" may be weakened and even somewhat dispersed, but it continues to "justify a relationship of inequality and the right to command."

Together, this attempts to describe the staying power of the paradox syndrome, as well as its ability to stay hidden. But let's not stop there. Thanks to Heather Cox Richardson and Immanuel Wallerstein, this is basically a 'rough cut' of an idea that I believe has transformative value. So, think about it. Discuss it with friends. Try your ideas out. That is the way we can together refine these ideas so that they work for us, we the people, and not just the Excluders and turn-about experts.

Earnestly Fooling Ourselves

We need to be clear that doubletalk also infects members of all groups; it's not just the turn-about experts who make use of it. This is true of communities and organizations, as well as most individuals, and it particularly typifies the political side of the paradox syndrome at the state and local level. In other words, it is not just national politicians or defense moguls acting out of their supposed self-interested political and economic rights. It is instead, multi-headed and includes much of society. For those that claim membership on the Excluder side, the modus operandi is a dual one: offers of privilege and self-interest appeal to multiple people who want to be like the Excluders and associate with them. Even on the Includer side—when they are attempting to 'do the right thing'— and not quite sure what that is, double-talk presents definitive pitfalls. And while compromise is necessary to solve many problems, the Fence-straddlers keep scrambling from one side to another irrespective of the problem being discussed simply because they like to *feel* that they are doing the right thing—and, of course, to assuage their driving self-interest. Thus, in more circumstances than we would like to admit, most of us—maybe all of us—have at some point suffered from a mild strain of double-talk.

Fence-straddler double-talk, while rarely recognized as such, is particularly popular. For that reason it is often used as a tool by the Excluders to deceive Fence-straddlers, enticing them to act against their own best interests. Double-talk is also well named because it promotes both individual and community deception, bringing together individual opinions with distorted cultural and societal systems of belief. In turn, this creates a double-edged tool that can be used to maintain negative difference and indifference, emphasized by fear of the "other" as well as a necessity to belong ... someplace.

The double-edged process of *fooling* the individual citizen so that they will continue to support traditional hierarchies and their actions—many of which are now unlawful—has been standard procedure for too long. In the Spring of 2020, this double-edged process was exemplified in the case of Breonna Taylor, killed at night, in her house, by police, during a home-invasion into the wrong house and address, using a police "no-knock warrant," later found to be illegal. Taylor, a medical worker, sadly fit all of the categories that would classify her as "less-than." She was a woman, she was black, and she lived on the West Side, labeled as the "wrong side of town." She was, therefore, an ideal candidate for injustice.

The police planned and executed the warrant with strikingly insufficient attention to Taylor's civil rights. They evidently didn't give them a thought. Federal prosecutors later described the Louisville Police Department's "Place-Based Investigations Unit" as "pursuing a reckless narcotics-trafficking case ... even though they knew a raid could place occupants of Taylor's apartment in danger." One officer pleaded guilty to federal charges of "falsifying a search warrant and trying to cover up the circumstances of what happened." Two other now former officers have been charged similarly. But no one has been charged with murder.[10]

Can we say the earlier, now forgotten permissions created by white community-belief-systems concerning negative difference, play a role? Do they somehow remain today? Yes, I would say so. While these permissions may be semi-forgotten, they remain active, even now, as operative social structures supporting negative thought-worlds. The illegal types of activities those police officers initially undertook that night exemplify the societal side of that double-edged ideological sword with its permissions to harm the 'less-than'. But this illegal action also depended on a community's individualized belief systems to support police actions, no matter what.

So, politicians are not the only ones who know how to tap into people's conscious individual beliefs and then pair them with their unconscious negative thought-worlds. Other groups are also highly adept at using these blinding individual ideologies to their advantage. In this situation, the Louisville police were skilled. As soon as the police's initial civil rights wrongdoing was revealed in Breonna Taylor's case, they immediately attempted to activate Louisville citizens' individual opinions against her and for themselves. The police may not have had any knowledge of subconscious thought-worlds, but they were adept at organizing community support based on these same thought-worlds. Therefore, they decided to characterize Taylor as a criminal and attempted to bribe a drug dealer in custody—a former boyfriend of Taylor's—to implicate her in his wrongdoing. When this man bravely refused, the police's case fell apart.

As a result, there was little of the planned for and expected citizen outcry against crime, supporting the police. The police equation method, "black drug-dealer boyfriend = black girlfriend must be a criminal too,"—failed and was laid bare for all of us to see, if we so wished. Some of us tend to shake heads in disbelief that individuals or communities would be so easily taken in. However, too often, these types of tactics continue to resonate with the negative thought-worlds hidden in our subconscious individual minds and cultural collectivities, then resonating with conscious cultural values and belief systems. And too often, we are not even aware of it because they are innocuously paired with opinions handed down from various venues that we trust—families, schools, churches, and universities. Thus, they are easily accepted even centuries later. As a result, intolerance, meanness, exploitation, and violence surely follow.

Self-Induced Confusion

Up to this point, we've been discussing how *group* subterfuge works based on their historical beginnings. But what about the individual? These processes of double-talk can also be individualized. Thus, it's also necessary to understand how these processes work personally, even for those who believe they are *truly* anti-racist, anti-sexist, anti-classist, anti-war, and definitively against any kind of inequality. But if those Includer beliefs and values are real and true, how would you, me, or any of us ever play a role in those negative processes? It's funny. Every so often, something can be learned from slightly stupid, awkward, and embarrassing circumstances—never-the-less related to more weighty situations. So, rather than insult someone else, let me skewer myself.

> *One day, my friend and I were standing in line at the local metro liquor store. I was picking up some last-minute wine for a Friday evening party at our house, and I had hoped to beat the crowd. But the store was already crowded, and the line was long. After 20 minutes when we got to the front of the line, I was ready with my drivers ID and checkbook (this was the 1990s). The cashier clerk—who I knew in passing—looked at me and said: "two pieces of photo ID please." I looked at him and responded, with some energy, but still politely: "What? Why? No-one asks for two pieces of ID." He said: "that's the rules."*
>
> *Oh, wow. I could feel myself warming up for a mild—still polite—tirade. All of a sudden, I felt a sharp rap between my shoulder-blades. My friend stage-whispered in my ear: "Paula, you're acting like a WHITE GIRL!" My spine snapped to attention; my eyes quickly went to the clerk. His eyes and face*

were rigidly neutral, but his lips were twitching. I knew he wanted to laugh as my friend and I were the only two white people in the store at the moment. I paused, let out a long sigh and said to him: "sorry, you're right" as I rummaged through my purse for that second piece of identification. As we left the liquor store, I looked back and smiled: he lifted an eyebrow and quizzically nodded.

Later, as I reflected on what had happened, I realized two things. First, at the beginning of the situation I was really getting into it with the possibility of a tirade—there was even a slight surge of energy. If I was honest, I felt slightly offended; I went there often enough that I assumed a <u>right</u> to be recognized—to have that rule overlooked—for me. I belatedly realized there was probably a sort of half-formed idea floating in the back of my mind; not so much as thought, but as background noise. The idea that I shouldn't have to give two pieces of ID; that in other businesses with a solidly middle-class clientele, if they recognized you, they didn't insist on that rule. So, why should the clerk here ask me, or anyone else in line that he recognized—white or black—for two pieces of ID? Wasn't this some sort of corporate discrimination? And you have to understand, in that 1990s timeframe, corporate misdeeds were in the news and much discussed.

But the idea that I was acting like a 'white girl'; that didn't seem quite right and hurt a bit. I didn't think my friend was right in saying that. We all knew—everyone in that store knew—even if the phrase "white privilege" was not yet in cultural use, that this privilege certainly existed. All we had to do was move that tableau just a little to the west, to a sister store about four miles away to a neighborhood that was primarily white. There, the black gentleman in front of me would have been asked for two pieces of ID, and I, if I so wished, could have gotten by with one. But it was Friday, we were going to have a party, so I didn't really take the time to think about it much. However, it did stick in my mind.

Looking back, I realized there was little, if any recognizable hurt that was incurred in that particular instance. So why was I thinking about it? Was I just trying to avoid the charge of 'white girl' in those circumstances? Maybe. But over time, I came to some conclusions that possibly explain my own intentions at that particular moment in more nefarious terms than I would like; and they might also have some difficult implications for making that move from wanna-be Includer to a true Includer.

To explain, I have always been one of those people who does not have much trouble speaking out when I witness something that I think is wrong. But here is the catch. That day, I claimed to myself that the reason I objected to the ID question was that I knew—we all knew—in nearby more white neighborhoods, they did not ask for the IDs of recognized customers. And to me, I told myself that demonstrated a discrimination against all of us—white and black—in that store that day. I was right about that. But upon some longer reflection, I also realized that when I raised that objection to the cashier, I had no real intent to follow up and do something about it.

So, if I am honest, I would have to say I was simply raising the question for my own convenience, my own self-interest. Yes, I did understand that corporations and businesses make inappropriate and unequal distinctions as to how they treated their customers. But I wasn't going to go home and do something about it. That day I was simply using my belief in justice for my own self-interest. I realized that a wanna-be Includer might act like that, but not a true Includer. In other words, being a true includer means avoiding every aspect, even the smallest, most minuscule part of acquiescence to inequality, and that includes fooling ourselves.

Thinking Differently

Expanding awareness sharpens consciousness of the unseen. In particular, it better explains how values and norms organize beliefs and actions in relation to our *conscious* life, while thought-worlds and unrecognized social legitimacies organize our *unconscious* life of acceptance, belonging, and definition of the true. This expanding awareness is not sufficient to take down and banish the intractable and immoral—existential and inter-connected—four ancient wrongs. But it definitely points out, once again, the path to be followed.

There have been numerous explanations and justifications for why injustice and inequality continue to exist. On the conservative side, the explanation is basic—it is believed to be human nature and therefore cannot be changed. From the progressive perspective, the explanation is that economic capitalism is the true culprit—it encourages greed and expands inhumanity. As a result, the moderates continually attempt to thread the needle, using their primary tools of political negotiation and compromise to align differences within the political-economic perspectives—but with little success except for their own self-interest. So, we are often led to believe once again, that this is "just the way the world is."

However, if we collectively utilize the critical insights learned from Achille Mbembe, we may begin to understand how to actually take down and banish the four ancient wrongs—Why? Because that distinction which Mbembe made is critically important: "levels of negative difference may vary, but these changes themselves make no difference in efficacy." Once again, I have to emphasize how show-stopping this is. It means that the concept of "negative difference," however diluted, still continues "*to justify a relationship of inequality and the right to command.*" And those turnabout experts use this concept of negative difference—sometimes watered way down, and now sometimes hyped up—over and over because it still works.

This identifies what needs to be changed, and we now recognize how our own actions sometimes get in the way. In other words, if there is to be success, we now know we must—acting as true Includers—dismantle 'negative difference and indifference' itself as a totality—not just its supposed piece-by-piece dilution. And to do that, we must teach ourselves, and each other, to recognize when even the slightest amount of negative difference is being offered to somehow convince us that some aspect or shred of inequality must still stand.

Notes

1 Thomas Piketty, translated by Arthur Goldhammer. *Capital and Ideology* (Cambridge: The Belknap Press of Harvard University Press, 2020), 1.
2 There are few numerical assessments of these types of situations, as further explained in Ch. 12. My estimations used here are based on 40 years of experience of observation and participatory organizing in more than 30 countries.
3 Etienne Balibar and Immanuel Wallerstein, *Race, Nation, and Class: Ambiguous Identities* (UK: Verso, 1991), 29.
4 Heather Cox Richardson, *How the South Won the Civil War, Oligarchy, Democracy, and the Continuing Fight for the Soul of America* (New York: Oxford University Press, 2020), xv.
5 Cox Richardson, *How the South Won the Civil War*, xv, xxix. The entire introduction deserves reading
6 Mark Danner, "The Grievance Artist", *New York Review of Books*, Nov. 2, 2023, 85.
7 Balibar, *Race, Nation, Class: Ambiguous Identities*, 29.
8 Marcus Weisgerber and Tara Copp, "Military Sounds Alarm at Proposal to hold 2022 Spending to last Year's Spending", *Defense One*, Jan 12, 2022, https://defenseone.com
  Mike Stone, "Biden to Seek more than $770 billion in 2023 Defense Budget", *Reuters*, Feb. 16, 2022, https://reuters.cpm
9 Connor O'Brien, "Senate Democrats Defied Biden in their Vote to Boost Pentagon Funding", *Politico*, July 23, 2021, https://politico.org
10 David Nakamura, "Trial to begin for ex-officer in case tied to Bronna Taylor", *Washington Post*, Oct. 30, 2023, https://washpost.com

Bibliography

Balibar, Etienne, and Immanuel Wallerstein. *Race, Nation, and Class: Ambiguous Identities*. UK: Verso, 1991.
Cox Richardson, Heather. *How the South Won the Civil War, Oligarchy, Democracy, and the Continuing Fight for the Soul of America*. New York: Oxford University Press, 2020.
Danner, Mark. "The Grievance Artist", *New York Review of Books,* Nov. 2, 2023, https://nyrb.com.
Nakamura, David. "Trial to begin for ex-officer in case tied to Breonna Taylor", *Washington Post*, Oct. 30, 2023, https://washingtonpost.com
O'Brien, Connor. "Senate Democrats Defied Biden in their Vote to Boost Pentagon Funding", *Politico*, July 2021, https://politico.com
Stone, Mike. "Biden to Seek more than $770 billion in 2023 Defense Budget" *Reuters*, Feb. 16, 2022, https://reuters.com
Weisgerber, Marcus and Tara Copp, "Military Sounds Alarm at Proposal to hold 2022 Spending to last Year's Spending", *Defense One*, Jan. 2022, https://defenseone.com

10
OUR WORLDVIEW NO LONGER FITS REALITY

Events come so fast now, one after another; we do not even have time to explain or rationalize them. Slowly, the answer is coming into focus. It's not bad luck, and it's not someone else's fault. The growing self-knowledge of ourselves as a society, acquired by reading about the four ancient wrongs, then analyzing them, changes our perspective. We come hesitantly and sadly to realize that it is our society itself that no longer fits with physical reality—think wildfires, warming oceans, and tornadoes. Nor does it fit a humane reality—think refugees, civilian casualties of war, poverty, unwarranted and violent killings.

But now our expanding awareness allows us to better understand the interweaving of our conscious life with the thought-worlds and social legitimacies of our unconscious and how things could shift for the better. But there is one more obstacle. In this chapter, to better understand how these sub-conscious thought-worlds endured, I outline the historical framework that kept them alive and well. History is always the context within which solutions can begin to be discerned. This story uses the motif of 'the King's men'—profiteers and believers-in-the-right-to-command—as sub-rosa representatives of the hold-over thought-worlds that have accompanied the 500-year slow change from monarchy's absolute rule to American democracy.

First Revolutions

Ever since that day in 1776, moving from our colony roots to an aspiring democratic country, we have consistently declared "liberty and equality for all." This ideal has been held high ever since as our nation's banner motif. Even though both 'liberty' and 'equality' were clearly limited to men—and white men at that—it was still a new and inspiring vision. Together "liberty and equality" introduced hope and aspiration for a new kind of life, living in a new world, creating a new creed, all of which came to be named democracy. So, what have we done since that moment? Until recently, we have not discussed it much. But from that moment in 1776 forward, a minority of people who believed in 'liberty or freedom'—but did not in any way agree with 'equality'—have done their best to forestall, undermine, and limit that new American worldview, and they have succeeded over and over—until now.

Sadly, that original vision of liberty *and* equality has never been truly achieved. This was not because it was impossible. It was because the majority, then and now, let the pushbacks against *equality* continue. They either contributed to the perverse actions of those 'King's Men,' as perennial Excluders, or they claimed innocence as perennial Fence-straddlers, looking away, refusing to recognize the destruction caused. Yes, equality is written in our documents, and when we speak of freedom and liberty, equality does follow—but it's always muted and in second place. Equality's legitimacy and "license to operate" was, and is, continually contested—in our courts, in our schools, and in our Congress. The truth is, our national worldview literally regards equality as a factor to be negotiated and compromised again and again, over and over.

This worldview, as discussed in the previous chapters, welcomes various sorts of negative thought-worlds. It is a worldview that allows truth to be corrupted and mistrust and immorality to grow and flourish. Those are harsh words. So, just for a moment, let's back up and think it through—starting with the historical difficulties surrounding the actual "constitutional" moment beginning the transition to democracy. Moving from the tyranny of the colony in America to the beginnings of democracy was never fast or easy nor was it easy in Europe, where America's historical era began.

In Europe, for more than 500 years, any revolt against the king and his nobles or the church had traditionally been crushingly punished. Thomas Piketty, in *Capital and Ideology*, tells a chilling story about a peasant revolt that broke out in 11th-century Normandy, France. The community had evidently held assembly meetings which were later arbitrarily judged to be a challenge to the power of the monarchy.

> Without waiting for orders, Count Raoul immediately took all the peasants into custody, had their hands and feet cut off, and returned them, powerless, to their families. From then on their relatives refrained from such acts, and the fear of enduring an even worse fate gave them still greater pause …. The peasants, educated by their experience, abandoned their assemblies and hastily returned to their plows.[1]

That level of arbitrary power allowing such barbaric cruelty has thankfully been banned by transition to democracies and even pseudo-democracies. But have you ever wondered, other than war, how such an extraordinary transition—from tyranny to democracy—was made? Actually, the answer is quite pertinent to our situation in this 2020 decade. Basically, during the 1600s and 1700s, English philosophers, along with Dutch activists who promoted 'economic capitalism' focused on the new ideas of free markets and the rights of men to own private property, decided to join with the French Enlightenment philosophers. Only then was it declared: "all men (sic) are created equal." Together, this group created a new *governing philosophy*, underwriting the democracy transition—messy, bloody, and unstable as it was.

Divisions of Right and Left

The American Revolution of 1775–1783 against the English monarchy, with its subsequent Constitution, began the first revolutionary transition against monarchy and its arbitrary power in the western world. Six years later, in 1789, the French declared its first National Assembly and drafted a Constitution also meant to free themselves from the centuries-old feudal era where the monarchy and the Church wielded power with absolute impunity. It is interesting to note that Marquis de Lafayette, responsible for overseeing the drafting of this 1789 National Assembly Constitution, was advised and assisted by Thomas Jefferson, then Ambassador to France.[2]

There is a particularly interesting story about that National Assembly that has endured over the centuries and connects to our situation today. They were a diverse group of representatives assembled that summer of 1789, gathering to draft the Constitution; there was, however, already division. The royalists, strong supporters of King Louis XVI, included not only the King's notables but also the King's profiteers—all together known as the 'King's men.' Upon entering the Hall that first day and wishing to separate themselves from the revolutionaries, or so-called "rabble," they gathered to the *right* of the presiding officer. When the anti-royalist revolutionaries entered, also wishing to make clear their separation from the royalists in both philosophy and action, they headed instead to the *left*. Since that National Assembly event, the left/right division has become shorthand for opposing political ideas, ideologies, and parties in European discussions, entering the American lexicon in the early 20th century.

These two opposing stances are more than an amusing story. They effectively mirror a root problem that faces every democracy at its beginning and at points of crisis. We hear a great deal about the transition from tyranny to revolution; a fair amount about the forthcoming constitutional documents, but much less about the 'governing philosophy' that evolves from these preliminary signature actions.

The new democracies of the European and American traditions necessarily abandoned the *radical political right* with their preference for continuing arbitrary power and inequality. They also necessarily abandoned the *radical political left* with their insistence on equality for the revolutionaries and all the people—often referred to as "the dangerous classes." Staying with either of these two oppositional sides, it was believed, would have destroyed any potential transition to democracy. On the left, the absence of any equality acknowledgement would clearly anger the street revolutionaries. On the right, the King's men were adamant that the "rabble" could not be trusted with power or governance.

Instead, the new democracies decided to go with a political *center*. This center group, named the "Liberals" with a capital L[i] as a stand-in word for liberty at that particular time, would not stand to the right with the King and those insisting on the necessity for absolute authority. Nor would they stand with the people's revolutionaries and their call for absolute equality. Thus, what was to become the moderate faction[3] stood in the center, enabling them to negotiate with both factions—a powerful position.

The 'King's men,' as the Radical Right, advocating for continued inequality, quickly acceded to the moderate faction. They felt they could still hold their own, understanding even then that seeming diminishment of their standing did not necessarily mean *loss* of power or profit. The Radical Left, already having won some of their freedom and equality issues, also faced their own version of reality and acquiesced to their status as a side-kick movement with the Liberals. Thus, center-positioning became the governance prototype for the new democracies.

In the United States, this 'moderate faction' regime traditionally includes both parties—Republican and Democrat. Until recently, this center-oriented political power has remarkably continued to hold the majority of clout for well over the past 200 plus years. This governance methodology to hold both political and economic power at the center absolutely necessitates 'compromise'—and there is nothing wrong with that. In fact, compromise is an honorable and absolutely necessary strategy to solve problems and govern effectively. But there is something very wrong, even immoral from a human and humane perspective, when the issue compromised is 'equality.'

---

i Capital L Liberalism began with the rejection of the monarchy in the 18th century, supports individualism, free laissez faire economic markets, and prefers little or no government intervention. Small l liberalism, beginning in the 20th century supports individualism, but advocates for civil liberties and social justice issues, and supports government intervention. Neo-liberalism and classical economics both strongly support a focus on economics but differ on how much government should be involved.

Immorality and Compromise

A question presents itself at this point. Was that perennial worm of potential negative compromise—residing within the moderate faction *center-governance strategy*—recognized immediately for what it represented? Most likely: people have a talent for quickly figuring out their own self-interest. Was it then welcomed? Maybe, maybe not. But in any case, early in US history, politicians—Democrats and Republicans together—unfortunately began the practice of negative compromise, *enforcing inequality* and thereby feeding the worm that would create distrust and eat away the integrity of their own powerful new democracy.

The review of the four ancient wrongs in Part I helps to answer the questions concerning "inequality" for both the past and present in ways not originally anticipated. While the Part I review clearly underscores well-known economic perspectives on inequality, results of the sprint methodology also delineate a particular political perspective, previously little noticed. I have to admit, I was surprised. My original intent in using the sprint methodology was simply to create sufficient background knowledge of the four ancient wrongs for effective use in deep participation—understanding their interrelatedness in order to create maximum impact. But the 'sprint' also provides a new and different insight as to 'how' and 'why' the four ancient wrongs manage to keep working. Not only at the deeper, profound level discussed previously by Achille Mbembe, but also working at current political and economic levels, despite what seems to be continuing best efforts for justice.

In brief, it is the real effect and endurance of a well-established but overlooked pattern of *inequality compromise*. These compromises were facilitated and made possible by that center-governance strategy decided upon at the very beginning of the US existence: they kept the possibility of an evolving true equality to a minimum. The first and most shameful 'inequality compromise' was, of course, that of slavery. But this same pattern goes well beyond that "original sin" and has managed to permeate the entire tapestry of our culture. One of these secondary "inequality compromises" took place early on in our country's history and set a precedent. It involved—what else—the "right to vote."

Benjamin Franklin was one of the first to claim *inequality* as a virtue in this regard. On the voting issue at least, he strongly believed there should be no right to vote "for those who had no property ... they are transient inhabitants." In 1787, his colleague, John Adams, went even further: "If the majority were to control all branches of government ... debts would be abolished first; taxes laid heavy on the rich, and not at all on others; and at last, a downright equal division of everything be demanded and voted."[4]

That first secondary "inequality compromise" way back in 1787 was a crucial misstep which began a continuing multitude of low-key, bi-partisan soft-ball compromises—John Adams ideas sound quite familiar, even today, don't they? In the summers of 2021 and 2022, Republican-run red states were following the same sentiments, if not the same procedures, as earlier expressed by their 18th-century colleagues. They offered new Bills to their state legislatures to diminish voting rights, recognizing that true majority rule emanating from full equality—without illegal limitations—would cause their future defeat and loss of power. There is a continuation here—a dynamic and a pattern—that cannot be disregarded.

But that is not all. It is the ruse of 'slow progress,' which we have been asked to believe in, which is the real kicker. This long-held belief says that given time, negative difference within our society will diminish and finally disappear. But the review of the four ancient wrongs clearly illustrates *the intended diminishment of injustice by slow progress never abolishes or completely suffocates negative difference and indifference—injustice is simply re-rooted.* Numerous incidents in the chapters on 'war,' 'poverty,' and 'environment' illustrate this 'slow-progress' as a multiple-opportunity strategy, with its capacity to distract the observer from its real intent. It turns out that the sprint methodology is particularly adept at illustrating these illuminating shifts. Focusing as it does on the inter-relatedness of the four ancient wrongs, it captures the actual re-rooting of injustice from one form to the next.

But wait, there is more. This same ruse of slow-progress is also counter-intuitively used as a sign of progress, justifying both our society's commitment to justice and its continuing inequalities. Here, the key phrase is: "our country stands for justice and we are making progress." In other words, 'easy-does-it,' 'go slow,' and a variety of other euphemisms are put in play to keep justice synonymous with slow—very slow—progress. This ruse of slow-progress is the mirror image of the matching bi-partisan center governance strategy, but its implications have never been fully understood. In other words, this chosen political façade—particularly for the white majority—has kept our inequalities in the closet for more than 200 plus years. It has operated within that chosen center-governance process all this time with little question.

Once said out loud, however, it can only be recognized as a cutting indictment of our long-standing center-governance strategy, which, by 2024, many of the Republicans had almost torn totally to shreds with their rush to the right. But in their turn toward authoritarianism, these Republicans also brought to light the underlying perspective of this center-governance strategy—"the principle of equality depends upon inequality" and its working corollary, "equality will destroy liberty." So, this strategy, which allowed that first 'inequality compromise' for voting rights in the time of Benjamin Franklin, has continued to give permission for similar permutations of power.

Think of the more recent inequality compromises always privileging power of the status quo: think of the slighting of Anita Hill's testimony concerning Clarence Thomas's nomination to the Supreme Court; think of Ollie North and his expansion of presidential executive privilege for war; think of the pardons given to the elder Bush's senior advisors; think of the expansion of the prison system by President Clinton; think of the now documented lies to carry on a 20-year war in Afghanistan; think of the seditious license given to then President Trump—and the list goes on. Each of these is extraordinarily diverse, rarely recognized, but similar in their negative impact as 'inequality compromises.' They were, and are, the "King's men" at work. Once formed politically, these negative compromises within a center-governance strategy have held steady—until now, the 2020 decade.

Return of the Shock Troops

Moving to the present and near future: 'the King's men' have once again returned, settled in with great force. They now promote chaos as a substitute to that center-governance strategy, using negative thought-worlds of fear and uncertainty with abandon. The last time the King's men arrived with such force was during the Civil War as Democrat segregationists. This time, however, they and their reinforcements arrived in the early 1970s as hard-right Republicans, an offshoot of the 1950s John Birch Society, and started their takeover of the entire Republican Party.

These same 'King's men' had cheerfully taken the backseat over 200 years ago in deference to the two-party moderate-faction and its center-governance strategy—believing that profits and authority would still accrue to them. Because of that stance, equality has been quietly denied over and over from the very beginning. When the Civil War arrived to challenge the inequality of slavery, it actually didn't bother them that much. As then members of the segregationist Democratic Party, they simply bribed the anti-slavery Republicans to remove Federal troops from the South after the Civil War—sadly, not much fighting was necessary with that group.[5]

That first time around, the 'King's men' also became the shock troops for the 19th-century Gilded Age. While hardcore segregationists gained the upper hand in the southern states and then filtered their ideology throughout the country, particularly in the West, this allowed Northern productivity to grow uninterrupted.[6] That was one reality. On the other hand, numerous black and voluntary organizations fought against these injustices and for new meanings of equality in the ensuing decades.[7] But those brave groups were finally defeated, or at least deflected, by the end of the 19th century. Once again, the King's men settled back comfortably into the still prevalent center-governance strategy. And they enjoyed this backseat of profit and authority until well into the 20th century.

Few of us remember it this way now, but the 1970s changed everything. Suddenly, national and international profits began a major decline; this made that comfortable but powerful backseat posture no longer viable for the Excluders. The worldwide economic recession of 1968–1970, now muted in memory for most of us, was for the King's men a clear warning that the world structure which their brand of capitalism was dependent upon was changing—and changing fast. This included four crisis-inducing losses: (i) the end of colonialism with its easy access to primary commodities; (ii) disappearing cheap labor; (iii) difficulty in externalizing production costs; and (iv) cuts in profit due to worker health and education benefits.[8] As a result, the King's men *economic* power was, for the very first time, in real danger.

In addition, with the passage of the Civil Rights Act of 1964, the King's men received a preliminary *political* warning. It unexpectedly became clear that a new brand of progressive politics was finally creating an enhanced awareness of inequality. There was an almost immediate realization that these changes would leave no place for the continuing operation of the low-key but substantial "right-to-command *political* power," which all the King's men had come to take for granted by way of those well-guarded inequality compromises. So, on both counts—economy and politics—the hard-right Republicans, as the latest representatives of the 'King's men,' decided to bail on the long-standing center-governance strategy.

As the Democratic Party began to put its segregationist roots aside, the King's men beefed up their encroachment on the middle-of-the-road Republican Party. Although still a minority group within the Party—controlling only the Goldwater faction—they decided to clearly communicate their complete disaffection with the loosening of the long-standing inequality rules. So, Richard Nixon, with his "Southern Strategy," was sent as a political messenger. The Southern Strategy's racist tactics clearly indicated that the King's men were no longer politically interested in participating in the 200 year Liberal-Liberty center-governing ideology if this new ideology was to become more accepting of equality. To our detriment, hardly any of us even came close to understanding that message.

Several years later, the King's men, now representing the hard right-wing of the Republican Party, sent their true emissary—their royal emissary. Ronald Reagan, with all his charming rhetoric of "shining cities on the hill," elected President of the United States in 1979, brought the real message that *government was the problem*, and his first initiative was deregulation for greater profit. The Reagan administration, in his first term, would cut taxes, cut welfare, deregulate markets, enhance financials of private property holdings, privatize and outsource public goods such as parks and postal services, while serving notice on emerging environmental legislation.

At the same time, in a low-key, crucial but less recognized methodology, Reagan politics focused on division and exclusion. Lee Atwater, President Reagan's advisor, in an interview in 1981 at the White House, clearly described this new direction in its most obvious and vicious terms. Most of us have heard or read this quote, but it's necessary to remember that it was not attributed to Atwater until 1991, ten years after his death and two years after Reagan and his administration had left the White House. Thus, the direct connection to Reagan's policy has rarely been definitively clarified.

But sitting in Atwater's White House office at that moment in 1981, the interviewer asks if President Reagan benefits from the racist vote by talking about "cutting taxes"—because the voter knows that's code for "cutting down on food stamps" or "doing away with legal services." Atwater replies with a stunningly clear response.[ii]

> Atwater: "Y'all don't quote me on this. You start out in 1954 by saying, "Nigger, nigger, nigger." By 1968, you can't say "nigger"—that hurts you. So you say stuff like forced busing, states' rights, and all that stuff. You're getting so abstract now [that] you're talking about cutting taxes, and all these things you're talking about are totally economic things ... But I'm saying that if it is getting that abstract, and that coded, that we are doing away with the racial problem one way or another ... obviously sitting around saying, "We want to cut this" is much more abstract than even the busing thing, and a hell of a lot more abstract than Nigger, nigger."[9]

From that point on, the Republicans played hardball to win. Congressional Democrats, a sizeable number of Congressional Republicans, and certainly, many of the American people, were, for the most part, all in the dark. No one—expert or not—seemed to understand that the *Liberal center*, with its 200-year-old agreements for 'inequality-compromises' had just been cancelled by a group of hard-right Republicans. Instead, the general assumption was that this was just one more economic strategy on the Republican side to win political votes. And it did capture votes; the Republican Party openly built new party allegiances by renovating old negative thought-worlds publicly justifying, as Martin Luther King had warned, of "racism, extreme materialism, and militarism." But no one truly comprehended the real and critical nature of this change by the hard-right King's men now based in the Republican Party.

None of Us Understood

The Democrats went on attempting to compromise in governance—losing ground every year—not understanding that the Republicans were no longer playing that old, established, but devious game of center-governance. For example, Clinton not only compromised, he fatally further facilitated the Reagan-Bush program on crime and poverty; Obama, believing in bipartisanship,

---

ii I state the quote exactly as it is so that the substantial perversion of this belief system is recognized.

also compromised in his political appointments, and compromised again, all with little to show for it. And then Trump, the 45th president, came along and said to all the King's men, in effect: "I like what you're doing, and I'm going to do it out loud!" Initially, many strategic-thinking Republicans didn't agree with that idea, but what could they do? As Trump's four-year administration continued, a number of these initial complainers were caught up in what can only be called the nihilistic spiritual zeitgeist of the moment, becoming followers and promoters of an anti-democratic white supremacist movement that promoted both lies and violence—all in the name of power.

So now we must *all* say it out loud: Yes, the Congressional Democrats fumbled and bumbled; and yes, they must have known that many of the corrupting 'inequality compromises' taken with their Republican colleagues over decades could never pass the smell test. But still, most of them did not understand that the capital 'L-Liberal' compromise agreement for center-governance itself was made null and void in the early 1970s by that current re-grouping of the King's men—the hard-right Republicans. Many of the House and Senate Republicans, however, did know from the beginning what was going on and grew accustomed to the attached power. That number has grown, particularly over the last decade—by 2024, it evidently includes almost every last one of them. And now there is no excuse for the Democrats to not understand, nor all of us for that matter. The center-governance strategy was, and is, dead-on-arrival. Instead, the King's men have moved to their preferred authoritarian stance.

So why is the fact that so few understood, so important? First, it caused extraordinary destabilization—and can, even now, still cause more. The now Republican 'King's men' had chosen an overtly racist and clearly anti-democratic path just at the moment—and exactly because—that 200-year-old capital 'L' Liberal-Liberty worm, with all of its inequality compromises and duplicity, was about to be fully recognized and changed. The 1964 Civil Rights and 1965 Voting Acts, passed under the tutelage of the Lyndon Baines Johnson (LBJ) administration, initiated real, substantial changes for true equality, focusing on voting rights and anti-poverty funding, for the first time since the Civil War. But the Republican Party clearly said no.

There is a critical background history behind all of this which the investigation of the four ancient wrongs—defining where we are, how we got here, and resulting understanding of the Paradox Syndrome with its built-in inequality—begins to reveal. Basically, in the 1930s President Roosevelt's New Deal began to nationally address economic inequality for the first time in US history. But Roosevelt allowed black people to continue to be specifically *excluded,* despite some diverse efforts to the contrary in his administration and in the country. Moving to the 1960s, in addition to the Voting and Civil Rights Acts, President Lyndon Johnson also signed the Economic Opportunities Act of 1965, acknowledging and addressing the poverty status of all poor Americans for the first time. In other words, *both the inequality of our previous political and economic laws and policies were officially addressed for the first time*!

These initiatives created critical long-term successes, but many others were stopped with great finality.[10] As a result, much of the potential success of this equality move was stopped because of Republican's shift to the right, as discussed. But most important—and this is what is neither recognized nor discussed—the subsequent Democratic administrations caved, particularly shifting to the right on economic issues for greater equality. Yes, they kept the words. But in the name of compromise they did not maintain that crucial policy and funding support to the working class and poor, as first initiated by Roosevelt, and then expanded to all groups, white and black by President Johnson. If the Democrats had recognized that moment, they could have more forcefully supported both economic and civil rights by also offering, at minimum, an inclusive economic justice bill—something that could have overcome the divisions within the country, and has yet to be passed.

Nancy Isenberg, author of *White Trash*, tellingly describes the situation from LBJ's perspective and why he believed the "War on Poverty" was so necessary: "As President, he never lost sight of how central class and race were to the fractured culture of the South." Isenberg observes that in private he was tellingly forthright in his criticism of all the players. About poor rural whites and the Republican Party, he proclaimed: "I'll tell you what's at the bottom of it. If you can convince the lowest white man he's better than the best colored man, he won't notice you're picking his pocket. Hell, give him somebody to look down on, and he'll empty his pockets for you."[11] Lee Atwater understood that, as does Donald J. Trump. And it explains, even today, why so many vote against their best collective interests.

LBJ used political insight to do the right thing in deciding to support and sign the Voting Rights legislation. In contrast, Ronald Reagan, Lee Atwater, and more recently Donald Trump intuitively used white people's fear of status change to further their Party's exclusionary interests. So, what did the Democrats do at that point? Rather than countering those Republican racist politics with an economic justice bill in the early 1970s, they chose instead a supposedly honorable strategy of emphasizing justice support for diverse minority groups. From a theoretical justice perspective, it seemed to be the right strategy; and in reality, it would have made excellent accompaniment to strong economic justice initiatives. But as a single political practice, it was a numbskull decision—without the awareness of the true historical context.

Those Democratic initiatives did increase minority justice to some extent, but not sufficiently. Because too many of their members belonged to that so-called 'moderate faction' and were not sufficiently interested, their compromises, over and over, watered down the impact. It was the same moderate factions of both parties, that earlier let economic justice and LBJ's

"War on Poverty" slide into oblivion. It was a critically bad and unjust decision—and Democrats did nothing to resuscitate it. If they had done so, Democrats could have effectively countered the re-alignment of working people, primarily southern whites, that is now Trump's Party, starting with the Carter administration. They also could have accomplished it with the Clinton and Obama administrations. But they did not. The Democrats kept the words of economic justice, but not the actions. Instead, they focused on racial and gender equality, but only mildly with many compromises—moderation in full sway.

The Root Cause

So, in this 2020 decade with change of one sort or another careening toward us, we must ask ourselves, we have to ask ourselves: despite our best efforts, how did this worldview with its ancient wrongs endure so long? And most importantly, how do we solve this seemingly unsolvable situation? Elections are critical, but they are insufficient on their own. We now understand that the Paradox Syndrome allows that previous colonial exclusionary template to be embraced and utilized wherever the power-holders wish. While it traditionally focused on blacks, people of color, and immigrants, the template's current direction in this 2020 decade has turned regressively backward to once again include women in the controlling grip of those long-enduring masculine systems. So, the question becomes: who's next?

This brings to the forefront the political realization that everyone is vulnerable to this "multi-sided crisis of inequality," not just those who have historically experienced the "less-than" categorization. Those political realizations are now making an impact on everyone. At the same time, the negative economic aspects of this crisis, something previously experienced only by working-class and poor people, is now a danger for middle and upper-middle classes. But this acknowledgement from Democrats has been slow in coming. As a result, multiple break-offs of political support for democracy have appeared, with increasing support for more authoritarian possibilities. However, in the summer of 2024, with the Democratic nomination of Kamala Harris for President, there was welcome political pushback to these Excluders and wanna-be authoritarians. But because this authoritarian resurgence continues to remain too close for comfort, let's reevaluate.

Several centuries ago, hierarchical "negative difference" was enforced by the brutal rules of rulers here and across the colonial world. But now, we have to ask ourselves, what enforces our current societal order? What is keeping that "inscription within a fundamentally inegalitarian and hierarchical order" which allows the four ancient wrongs to remain operative and in place here in the United States? It is neither the military nor the economic and political power holders who, while they may take advantage of the existing social order for profit and power, they do not enforce it. Even they do not have the right or sufficient means to physically enforce that exclusionary inequality inscription, as long as we remain a democracy.

The only tenable conclusion is that the four ancient wrongs keep their strength and that the "distinct institutional system" with its fundamentally inegalitarian and hierarchical order keeps operating because it remains *socially accepted*. This disturbing conclusion has been missed time and time again. And it's primarily because we just don't want to see it. What enforces the system itself and diverts efforts to dismantle its inequalities are neither economic nor political. Certainly, the political and economic actions contribute to the ease of our blindness, but they cannot enforce it. It is enforced only by our ongoing *social acceptance*.

This enforcement is held within the social order—not the economic order, not the political order, but the social order. And the divisions separating political and economic equality contribute to the ease of our willful blindness. So, despite the political/economic methods utilized to reform the system, the four ancient wrongs retain their strength, as does the social order itself. It will continue to do so until it is no longer socially accepted.

That is the bad news—now the good news. Because it is 'socially accepted' we can easily change this situation, even though it has existed for more than 250 years. All we have to do is say: "No, we no longer accept that." There are no wars to be fought, no further trauma to be suffered. We just collectively change our minds, and here come the needed culture shifts and social transformations. But, of course, to effectively and collectively change our minds, we must repair our minds. So, there is still some work to be done.

Notes

1 Thomas Piketty, *Capital and Ideology* (Cambridge: The Belknap Press of Harvard University Press, 2020), 66.
2 Britannica, "Classical Liberalism", https://britanica.com. This essay emphasizes the necessary key points to understand the longevity of this political doctrine
3 I use this 'moderate faction' term to distinguish from current political nomenclature. This term as applied historically would include the majority of members from both political parties. Today, however, with the alt-right takeover of the Republican Party, the few moderates, if any, who exist in the current Republican Party hide themselves for fear of losing their next election.
4 Britannica, "Classical Liberalism", https://britanica.com
5 Eric Foner, *The Second Founding: How the Civil War and Reconstruction Remade the Constitution* (New York: W.W. Norton, 2019), chapter 3: The Right to Vote, 55–92.
6 Kevin Waite, *West of Slavery: The Southern Dream of a Transcontinental Empire* (University of North Carolina Press, 2021).
7 Kidada E. Williams, *I Saw Death Coming: A History of Terror and Survival in the War Against Reconstruction* (New York: Bloomsbury Press, 2023).
8 Immanuel Wallerstein, *The End of the World as We Know It* (Minneapolis: University of Minnesota Press, 1999), 30–32.

9 Rick Perlstein, "Exclusive: Lee Atwater's infamous Interview on the Southern Strategy", *The Nation*, 2012. The actual recording of the interview is included in the article, https://thenation.com
10 Ryan Larachelle, "A Mission Without Precedent", *Journal of Policy History* (Cambridge University Press, 2024), no. 1.
11 Nancy Isenberg, *White Trash: The 400 Year Untold History of Class in America* (New York: Penguin House, 2016), 264.

Bibliography

Encyclopedia Britannica. "Classical Liberalism", https://britanica.com
Foner, Eric. *The Second Founding: How the Civil War and Reconstruction Remade the Constitution*, New York: W.W. Norton, 2019.
Isenberg, Nancy. *White Trash: The 400 Year Untold History of Class in America*, New York: Penguin House, 2016.
Larachelle, Ryan. "A Mission Without Precedent", *Journal of Policy History*, Cambridge University Press, 2024.
Perlstein, Rick. "Exclusive: Lee Atwater's infamous Interview on the Southern Strategy", *The Nation*, November 2012.
Piketty, Thomas. *Capital and Ideology*, Cambridge: The Belknap Press of Harvard University Press, 2020.
Waite, Kevin. *West of Slavery: The Southern Dream of a Transcontinental Empire*, University of North Carolina Press, 2021.
Wallerstein, Immanuel. *The End of the World as We Know It*, Minneapolis: University of Minnesota Press, 1999.
Williams, Kidada E. *I Saw Death Coming: A History of Terror and Survival in the War Against Reconstruction*, New York: Bloomsbury Press, 2023.

11
SOLVING THE UNSOLVABLES

Achille Mbembe, the author who helped us understand how *negative difference* was used to inscribe a fundamentally inegalitarian and hierarchical order in our society at its beginning and in its continuance, has also memorably written: "The world will not survive unless humanity devotes itself to the task of sustaining what can be called the *reservoirs of life*. The refusal to perish may yet turn us into historical beings and make it possible for the world to be a world. But our vocation to survive depends on making the desire for life the cornerstone of a new way of thinking about politics and culture."[1]

Becoming historical beings, making it possible for the world to be a world, building a cornerstone for life dedicated to both equality and freedom, adopting a new way of thinking, all begins with divesting ourselves, our culture, our society, from that long enduring *inequality* underpinning the four ancient wrongs. Now we begin to understand that we can actually eradicate this intertwined inequality by using the same social power that has protected it. Thus, the reality of a new situation slowly clarifies itself. As Thomas Paine said, we begin to "see with other eyes" and "hear with other ears."

Crumbling Justifications

Our world changes with the recognition that it is social acceptance which continues to enforce the right to practice inequality and not the political and economic powers we have preferred to blame. Recognition at first is slow in coming because we concentrate on what has been done in the past rather than what can be done in the present. However, the power to say "no" to the ongoing practice of the four ancient wrongs and their inherent inequality, because of its social placement, actually rests with all of us—not anyone else. We don't have to fight; we don't have to tear apart our society. Realizing that it is only our continuing, unconscious social acceptance of inequality that *enforces* this intertwined, long-enduring practice, we can decide to cease and desist. Then we can determinedly begin to identify both the obvious and hidden mechanisms that maintain the inequality practices in place. With that process ongoing, dismantlement of the four ancient wrongs begins at the same time as the old justifications for inequality crumble apart. All we have to do is collectively change our minds—the definition of a culture shift.

Certainly, the power-hungry and the profit-driven will attempt to obstruct. In fact, extremist forces are already doing so. But these various groups distributed throughout the country are simply and sadly the pawns for the Excluders and economic profiteers who wish to maintain their long-enduring *equality* based on *inequality*. These groups and their supporters may present themselves more civilly than those that stormed the Congress on that infamous Jan 6 day, but they are no better. While they have taken advantage of this long-enduring inequality practice and want to continue doing so, they have no legitimate power to enforce it. Yes, they have some *political* power, but they do not have the *social* power that remains with all of us. In other words, the power to peacefully reinstate one of our society's major organizing principles—equality—remains almost exclusively with us, the people.

Remember, however, Thomas Piketty told us that unless justifications for society's inequalities are found, "the whole political and social edifice stands in danger of collapse." We as a country are now unthinkingly careening down the road toward that point; the contradictions and chaotic conditions are there for all of us to see. Thus, we need to prepare a new societal 'edifice,' throwing aside the elements of the old inequality practices and yet keeping and preserving the best of our society. If we are not so prepared, the Excluders will take over as they attempt to move in with further chaos. That is what happens when instability increases, and trust in society, its government, and its institutions all diminish. That is what is happening now because the Excluders recognize that all of their necessary justifications for inequality are crumbling. Therefore, their game is to introduce further chaos to heighten greater acceptance of strong-man authoritarianism or, if need be, move on to violence. For Excluders, chaos fits them just fine.

So, how do we do this? How do we create that new and better societal edifice? This question is not about competitive economics or the politics of diverse identities which the proponents of inequality use in their attempts to distract us. It is not about any of the normal, everyday cyclical political and economic changes that stable societies experience. No, instead, it is about what unstable societies and communities experience as the crumbling processes of demise begin. But we have the alternative to hopefully move toward a new genesis—a new beginning. In other words, it is a critical change-point, and we have a choice. We can realize that it is about us as humans, the nature of social relations and our basic human morality, our connection to the protective planetary life that surrounds us, and the society we choose to create and live within, or not.

There is, however, a particular importance to the ramifications of such a change-point which needs further emphasis. It requires this further emphasis because current sociology inquiry and social theorists, as discussed in Chapter 8, still remain primarily concerned with the problems of a stable society. As a result, they tend to ignore the dynamics of social instability and their implications. The critical role of such change points was, however, identified back in the 20th century by A. R. Radcliffe-Brown, a social anthropologist known for his combination of theory and practice. Paraphrased for better understanding, he counsels the following:

> I would suggest that we call the first kind *readjustment*. Fundamentally it is a readjustment of the equilibrium of a social structure. The second I would prefer to call *change of type*. However slight the latter may be, it is a change such that when there is sufficient of it, the society passes from one type to another.[2]

Radcliffe Brown observes that "it is absolutely necessary to distinguish and study separately" the two types. That is particularly good advice for all of us. The specificity concerning different kinds of change is important to understand because the societal edifice change proposed here is a change-of-type. It is not simply a readjustment to the existing societal structure. Instead, it is a type-of-change where the society itself changes its structure.

In other words, readjustment change, which has worked effectively in a stable society with its implicit competitive/conflict dynamics, does not work in this change-of-type situation. Instead, finding the best, shared-knowledge response will demand a socially oriented solidarity base, featuring both deep participation practice and social integrative power, which is not normally required in everyday participation. In a democratic country, these more profound ensuing collective agreements will take some time. But in the interim—right now—the just completed investigation of the four ancient wrongs, with the review of their consequences and our present status, gives us an effective place to start.

There are two resets of our current social edifice, encompassing the political/economic, which can kickstart this new and necessary process. We need first, to reinvent our mental imagery to create a culture shift that results in acceptance of non-negotiable equality, and adoption of social relations as a first organizing principle. If we do so, we will have the space to sustain those "reservoirs of life" as well as the ability to construct those necessary "living forms of solidarity" both comfortably ensconced in a more trusting, and trustworthy, society. It is within creating this process that we can solve the unsolvables and perhaps become those historical beings we would like to be, capable of laying that new cornerstone of life.

Deciding to go forward in this direction emphasizes a critical point to keep in mind. If we are to remove the social acceptance of inequality and its practice as the root cause, it requires that we focus on the inequality acts and processes themselves, not simply on the people doing the acts. This is where the difference—highlighted in the Introduction between *righteous anger* and *self-righteous anger*—becomes important, particularly for those of us who wish to be effective Includers. If righteous anger against injustice is self-appropriated, it becomes *self*-righteous, creating dynamics that are polarizing, divisive, and self-defeating.

Instead, what works are initiatives that promote solidarity, social energy, and collective critical-thought initiatives. In light of this preferred orientation, there are also effective defensive methods to be considered that take into account the ongoing efforts of the Excluders to maintain their current status. As these defensive methods become more automatic on the Includer side, the positive energy *reform, reinvention, and reimaging* methods of deep participation and social integrative power begin to create positive impact. It is these trust-based methods which are most effective in culture-shift and mind changing. But the defensive methods are also necessary.

Disabling Thought-World Camouflage

A review of defensive methods helps to begin the disabling of negative camouflage and stabilize the foundation for positive societal change. Negative thought-worlds, those wily, slippery creatures that hide in the recesses of our sub-conscious, are best described as "reptiles of the mind."[3] The role they have played in keeping transformative change always just out of reach is one we need to continually keep in mind. As awareness grows that the major camouflage support of inequality—social acceptance—is now recognized and publically apparent, negative thought-worlds will still continue to be used, just differently. This time, Excluders will play on negative thought-worlds to create confusion that *blocks*, in various ways, those social acceptance realities from spreading and becoming common knowledge. To effectively negate that process, it makes sense to first understand how negative thought-worlds interact with the notions of inequality, negative difference, and indifference.

"Thought-worlds" are a shorthand way of referring to the role society plays in organizing social thought for the individual. Most important, the thought-worlds themselves—positive, negative, or neutral—remain out of sight but provide cover and give legitimacy to the *implicit* rules of culture and society. In turn, these same implicit rules govern the *explicit* rules and values of society's social, economic, and political world. Well-anchored in both our individual and collective sub-conscious, thought-worlds become more apparent in times of increasing political and economic instability. At that point, they become not only more easily recognized but also more easily addressed. But always residing at the sub-conscious level, thought-worlds remain, for the most part, unexamined in terms of either truth or basic human morality.

As a result, thought-world camouflage retains tremendous power. When negative in substance, for example, it fosters a societal culture that at least accepts and often practices acts of injustice and violence. In this situation, the camouflaged ancient wrongs are so often strongly built into economic, political, and societal structures that they are able to disappear into the background and exist without note. These negative thought-worlds certainly cover cumulative behavioral missteps passed down from one generation to the next, as illustrated, for example, by the 'masculine systems' of wealthy and powerful men. But the thought-worlds themselves should never be consigned to some kind of multi-generational conspiracy. Instead, it is their semi-conscious nature and constant repetition—energized by *social acceptance* and *acquiescence*—that keep them alive.

This foundational negative thought-world of social acceptance and acquiescence, we now understand, clearly supports the root cause of inequality practice and inter-relates the injustices of racism, sexism, and class with endless war, poverty, and environmental plunder. But three other negative thought-worlds—*hierarchy, self-interest, and distrust of government*—also support this negative foundational base. For example, belief in a top-down '*hierarchy*' based on domination provides a social order where negative difference and indifference are morally acceptable. Belief in the primacy of '*self-interest*,' supported by assumptions of 'scarcity,' provides a rationale for selfish hyper-individualism exemplified in the saying: 'if you gain, I lose.' *Distrust of government* provides a rationale for rejection of any government authority to redistribute property from individual to community and legitimizes violence against government. But keep in mind, none of these thought-worlds are conscious; instead, they subconsciously *shape* conscious ideas.

There are two basic methods most often used by Excluders to guarantee the ideas and actions they are suggesting resonate with the negative thought-world miasma existing in almost everyone's subconscious. Because there are only two, the defense against activating these negative thought-worlds becomes easier. The methods that Excluders normally use are *emotion*-based disinformation to introduce fear and anger or *intellect*-based disinformation to suggest superiority and importance.[4]

Emotion-based initiatives usually use a short phrase designed to have "emotional wallop" going directly to a person's anxiety, fear, and their sub-conscious beliefs in negative difference. The second intellectually-based method is designed to flatter a person's intellect, coercing conscious intellectual beliefs to be congruent with the underlying inequality premises through disinformation and misinformation. Both ultimately play into support of negative difference and indifference.

A good example of anger and fear is found in the current use of "anti-woke ideologies." Some people, holders of relevant negative thought-worlds, and often prompted by Excluder double-talk, promote the idea that citizens who are *different*—people of color, recent immigrants—for example, should not be allowed to vote because they are not true Americans (read white). Even more bizarre, these conscious ideologies, once in place, supported by sub-conscious negative thought-worlds, promote the idea that these same people, because they are ostensibly different, may also be dangerous and want to hurt children.

Other intellectual strategies might feature claiming that diverse groups are introducing harmful ideas or feelings into classrooms, or harmful books into school libraries. The Excluder claim is that these actions induce feelings of shame in our country's history or to indoctrinate and groom students to become immoral by introducing them to gender differences. These ideas are best described by most of us as cringe-worthy attitudes, but they are not so to all;. Instead, for these groups, their sub-conscious fears and anxieties, supported by their unidentified but very real negative thought-worlds, are exploited by Excluders to maintain an ongoing negative difference.

Obviously, these initiatives are not attempts to improve our educational system through reasonable dialogue. They have only one objective: to strike fear and anger in parents so they will support muzzling basic critical thought in the classroom by limiting the teaching of history and its challenging ideas—hard to believe, but it works. By doing so, Excluders narrow the public space for democracy to remain vital and decrease freedom by labeling particular ideas and people as malevolent and dangerous. An example of outcome, the Stop W.O.K.E. Act, was passed in the state of Florida in 2022 (now under legal appeal) and prohibits instruction on history and current events of race, diversity, and sexism in both the school and workplace.[5]

Other techniques directly provoke negative difference and indifference. Some of the most prolific of these techniques are zero-sum games—*if you win I lose*—strongly attached to self-interest *and* indifference. These are endemic, and they often evoke immediate emotional responses—a good sign of sub-conscious thought-worlds at work. They include being against adequate public funding for childcare, worker compensation, physical health care, mental healthcare, access to housing, and education, among others—all examples of the "if you win, I lose" mentality. One of the more recent zero-sum stand-off games was played against President Biden's executive action student-debt reduction plan in 2022 on student loans totaling $1.6 trillion. The *complaints* against this reduction are multiple: "that's not fair to those who paid their debt"; "the country cannot afford it," and the multiple complaints continue.

But in reality, it's that self-interest, that early-on zero-sum learned assumption speaking once again from the sub-conscious—if you win, I lose.

The real facts remain sidelined in zero-sum games such as these. For example, we are still a dominant player in graduate education with 40 US universities ranked within the world's *top 100*. But if the facts of the situation are checked, we will find only 23% of US colleges are ranked within the *top 1,000* undergraduate universities worldwide. Why is this? A columnist for "Best Colleges" says it well: "American higher education is delivering for the elite, but not for the masses."[6] In other words, state universities that focus on undergraduate education, once seen as a *public-good* primarily supported by state taxes, and of great benefit to everyone, no longer serve that need. Instead, public universities which remain of particular benefit to the middle and working classes who often cannot afford private university tuition rates but still deserve and want a university education have seen their tax-support slashed. They are now seen as a *quasi-private-good* supported by individual students forced to pay high tuition rates. If you have family wealth, that works, but what if you don't?

In effect, each one of those people carrying large amounts of student debt is evidently considered "less-than" by the representatives who sit within the halls of political and economic power but do not support public universities and their students. These representatives promote zero-sum games possibly because that is their preference. But it is also because the majority of us continue to prefer, or at least tolerate, unexamined thought-worlds. In these situations, *self-interest and indifference* will win the day, as opposed to consideration of the needs of others. Heather McGhee's compelling book, *The Sum of Us: What Racism Costs Everyone* explores the damage done by scarcity's underlying zero-sum-paradigm to *all* of us—"the idea that progress for some of us must come at the expense of others."[7]

All of the above negative difference techniques are explicitly designed to elicit some level of either anger or fear. A second method that brings difference and indifference into play blocks intellectual recognition of the inequality structures and systems themselves. It does so by highlighting a supposed positive *hierarchy* and playing to the supposed intellectual proficiencies of the reader. Recently, I read a column written by a longstanding, respected reporter of the Washington Post, complaining about the "equity agendas" of progressives who were attempting to close the achievement gap in schools. This columnist claimed that for the progressives involved, "the antonym of equitable society is a meritocracy."

OK, let's stop here for a moment: do you know what that previous sentence means? I wasn't positive: I had to look up the meaning of 'antonym' to be sure I understood what the columnist contended. And yes, I know that we all learned about synonyms, homonyms, and antonyms in 7th grade English, so it is not an intellectual word used only by brainy geniuses. But it is obscure language, used to suggest to the reader they are reading something at least slightly profound, perhaps available only to those of higher intellect—and so the reader is possibly coerced by flattery to believe the substance of the article.

Now, I don't know if the actions of the progressives in question were effective or inept. But I do know that the progressives themselves did not say what the columnist stated; that was the reporter's interpretation. Evidently, the writer either knowingly or unknowingly, was working from the assumption that minorities are inferior and therefore used governmental rules to gain advantage. The columnist then gave these ideas of 'negative difference' a slight intellectual sheen, offering a gloss of high IQ by inferring to the readers their membership in an implicit intellectual *hierarchy*. However, by relying on and even promoting hierarchy and negative difference, the columnist does the dirty, profiteering work of the Excluder by inferring inferiority and, at the same time, thoughtlessly diminishing trust in government. Perhaps he did not have this intention, but unconscious reliance on these negative thought-worlds is how they work.[8] Recognizing these Excluder methods to bring negative thought-worlds into play makes pushing them aside and paying no attention much easier.

The Negative in the Positive

To live a happy life, people are often advised to look for and recognize the positive. However, in defending ourselves and others against the vagaries of negative thought-worlds, the best advice is exactly the opposite—it is the negative lodged in the positive that is dangerous. This is particularly true for negative camouflage activity, which keeps inequality hidden and operating, even for those who believe in equality.

When *'self-interest'* is paired, for example, with the idea of 'growth,' it has the ability to evoke positive associations. While ecological destruction of the planet is now identified as one of the biggest threats to humankind, its danger continues to be easily dismissed by using 'growth and prosperity' phrases; pleasingly designed to appeal to, yes, well-being—but also to self-interest. Because this phrase, 'growth and prosperity," and others like it underwrite our economic foundation, too few of us even consider that "growth" can even be the true cause of our environmental crisis.

Delinking the positive word of 'prosperity' from 'growth' begins to take away the cover used by those who are and have been, from our country's beginning, possessed with the necessity for possession. However, even if that positive association is taken away, a negative assertion can still kick in as a truth, operating on both positive and negative wavelengths. In this situation, "scarcity and self-interest" is explained as the quintessential "fight for survival," subliminally associated with our mythic frontier ancestors, thereby legitimizing violent methods of aggressive and acquisitive dispossession. Hewing to our mythic frontier past works well for any business corporation that destroys something to make something. Their efforts to dominate and profit while disguising their actions, no matter the danger to human and planetary well-being, are widely practiced (think actions of petroleum, big pharma, and tobacco corporations).

Distrust of government was also glimpsed throughout the four ancient wrongs in various times and situations, but its amplification as a supposed positive began when President Reagan came into office in the 1980s and immediately announced, "government is the problem." This set the stage for a new era that recycled old messages about government welfare emphasizing negative difference in terms of the poor (think "welfare queens," a term which Reagan introduced) or of asylum seekers and those seeking to immigrate (which Trump savaged with multiple negative terms). These sound bites, generating distrust, were aimed at the federal government and their actions to build an equitable and just national community. At the same time, these same negative statements were often regarded as reliable because many people believed the President was honestly warning them of danger. Repeated again and again, these fear-mongering messages began a 40 plus year deluge of distrust against public institutions. Reagan Republicans then, and MAGA Republicans now, love those messages; they are strongly aligned, subconsciously, with that working corollary: "equality destroys liberty."

Bill McKibben, the inveterate environmentalist who never gives up, sums all of this hate-filled negativity up, tongue-in-cheek. He suggests that it might be Ayn Rand, the novelist, who is "the most important political philosopher of our time." He then points out that this idea is, of course, nonsense. "Her ideas about the world are simple-minded, one dimensional, and poisonous. But you don't have to be right to be influential." McKibben goes on to remind us that Ayn Rand has had, and continues to have, enormous political influence. As he explains it, there is a reason that those influential years of Margaret Thatcher-Ronald Reagan could be characterized as "the second age of Rand … when the laissez-faire philosophy went from the crankish obsession of right-wing economists to the governing credo of Anglo-American capitalism."

This focus on Rand's "raw attitudes," according to McKibben, packs that "emotional wallop." Her phrases, all derivatives of the basic negative thought-worlds, include: "government is bad"; "selfishness is good"; "watch out for yourself"; "solidarity is a trap"; "taxes are theft"; "you're not the boss of me."[9] Now, with increasing virulence, that "emotional wallop" continues to expand, and Ayn Rand remains the favorite reading material for MAGA Congressional Republicans. In June 2023, for example, a Republican Congressman from North Carolina explained his flip-flop by first insisting, during a debt ceiling standoff, that the Government could not afford to pay its bills. Two weeks later, the Republican congressman voted for a bill that would add more than $3 billion to the federal debt. Asked to explain, he said he had been reading *Atlas Shrugged*, Ayn Rand's well-known sermon on self-interest, and had learned "a wonderful lesson."[10]

These are all reasons to start paying attention to how the four ancient wrongs and their supporting negative thought-worlds are defacing the communities we love and defaming our nation. Surfacing these sub-conscious organizers of shared social thought is a primary way to begin the necessary dismantlement of the ancient wrongs and, with that, the "inequality principle" around which our society has been sub-consciously organized. By understanding the negative intent, we can reject and help others reject the intended fear and anger. We can then begin to focus our energy on both the living beings and the planetary world around us, which both require manifestations of equality and reciprocal care.

But again, let's be clear about what we got wrong. Yes, the physical realities of the four ancient wrongs are multiple in our politics and economy and underwritten by the rarely recognized inequality principle. But it is the collective mental thought-worlds that keep them going. More specifically, it is the *social acceptance*, with its accompanying acquiescence, generated by negative thought-worlds, which keeps the four ancient wrongs in existence.

Enduring Complicities

Every good working system has strong backups at the ready. And Achille Mbembe's "distinct institutional system," housing that distinct American principle, "equality depends upon inequality," is a strong system. Because awareness of the four ancient wrongs has been slowly expanding for the past sixty years and clearly accelerating over the past two decades, it is no surprise that more conscious backup systems supporting thought-world camouflage have also developed. A key backup, the practice of complicity, is essential for politicians, economic profiteers, and their followers—all the Excluders—who wish to keep this inequality system up and going.

A dialogue between two well-known writers, Anand Giridharadas and Chiara Cordelli, in Giridharadas's book, *Winners Take All,* explores the idea of complicity in ways that support the investigative results of the four ancient wrongs. In particular, the two writers point to the role that *acceptance* plays in this wrongdoing. First, they make the distinction between "active committers" and "passive permitters" of harm. Most of us have no problem identifying those active committers. If someone doesn't pay taxes while making millions or runs a hedge fund that puts profit before people; all of these are the "easy cases" illustrating "direct complicity, consciously practiced." Then there are those of us—perhaps most of us—who live more ordinary lives but still attempt to alleviate suffering to make the world more just, who may still suffer from some level of often unintended, unconsciously practiced complicity.[11]

If, for example, I am a local volunteer with an environmental group advocating for green energy, but at the same time I am also invested in mutual funds that have small holdings in Exxon Mobil and thus benefit financially as part of my retirement fund, I am, according to this definition of complicity, a "passive committer." In other words, neither I nor you have to be an "active committer" to be complicit. Even when well separated from the injustice committed, if you or I possibly profit, we become passive committers of harm. Cordelli calls these the "harder cases," labeling them the "passive permitters," committing two different types of acts: "*alongside the act of helping was a parallel act of acceptance*," one moral and one immoral.[12] This is admittedly harsh, but when we think it through, it is also admittedly true. In other words, complicity is one more way of acting as a Fence-straddler.

Economic complicity, always endemic, is now expanding in ways that can involve those intending to do good in actual nefarious activities. Many charities, think tanks, and academic programs have recently adopted private/public 'fix-it mentalities,' based on win-win philosophies as documented in Giridharadas's book. Increasingly, these groups use short-term business values for "doing good while profiting from the status quo." For example, well-known think tanks hold conferences on how to 'fix' the environment. But the fixes turn out to be cosmetic, while the profit continues to flow.[13] There is an interesting story told by Herman Daly, the eminent ecological economist who figures in the previous environmental crisis chapter, which illustrates how these business values are adopted.

Daly tells us about his involvement with a Woodlands, Texas, Conference series sponsored by George Michell, a well-known oil mogul who was inspired by the book, *Limits to Growth,* written by Dennis and Donella Meadow in the early 1970s. The first Woodlands Conference series, held in 1975, featuring both pro-growth and anti-growth advocates, was, according to Daly, "a successful beginning." All sides had a respectful hearing, and four more conferences were scheduled. But then something happened. Daly explains:

> Somehow by the third conference the theme had mutated, from 'limits and alternatives to growth' to 'management of sustainable growth.' As a result, the leadership passed from Meadow and Meadow to the Aspen Institute and the University of Houston; and instead of challenging business as usual, the emphasis shifted to sucking up as usual to business interests. The new 'more balanced' view was that we really must not limit growth, just focus on good growth rather than bad growth. Growth had somehow become 'sustainable,' contrary to the main conclusion of *The Limits to Growth*.[14]

When I recently read this story by Daly, I was struck by the similarities that Daly and Anand Giridharadas were separately exploring about the small but impactful complicity of fence-straddling. Daly's experience was a singular series of events, while Giridharadas discusses numerous events, including several sponsored by the Aspen Institute and others like it. But both document small but critical change in emphasis. Each one illustrates how the focus moves from challenging unlimited business interests to, if not 'sucking up,' at least facilitating these interests (and yes, the eminent Professor Daly really did describe it as "sucking up"). Daly's story is, at one level, just one more illustration of how that *inequality principle* is kept in place to continue profiting from environmental plunder. But it also vividly illustrates the almost invisible but very real damage done by economic complicities.

These backups to the long-enduring camouflage of negative thought-worlds involve far too many of us in unremarked political and economic systems of complicity and fence-straddling. These nefarious systems are easy to miss, but when recognized are fairly easy to take down. So now we understand these activities for what they are—efforts to keep the inequality principle up and running so that both the status of power and wealth remain with the dominating primarily white majority, where it supposedly should be. Now, with new complicity recognitions in place, peaceful takedowns can commence. In these dismantling situations, moderation and compromise, the hallmarks of small complicities and fence-straddling, have no place.

Moving Beyond Politics

Now reality finally hits home. Fear of exposing the inequality principle and all its machinations has remained the true third rail of our time. Inequality practice has been interwoven so tightly among and between us all that most can't see it; or, shall we say, we don't want to see it. It has been so from the beginning, and it remains so today. Yes, people like Dr. Nicole Hannah-Jones, New York Times Pulitzer Prize winner, or Dr. Eddie Glaude, professor of African-American Studies at Princeton, speak to this inequality principle on MSNBC. They not only touch that third rail, they grasp it when invited on cable news and clearly point out that as a country we currently have two choices: accept that we are a multi-cultural, multi-racial country; or admit that we are a white supremacist country and wish to remain one.

At that moment, everyone sagely nods their heads in agreement but immediately says to themselves: "well, that's not me; I'm not a white supremacist." However, those background self-deceptions are still hard-at-work and they easily continue because Jones and Glaude are asked to speak only of racism, allowing all of us to miss the statement's broader import. Let's stop for a moment and think this through. Politically speaking, yes, one could say the claim that we are a "multi-cultural, multi-racial country" is a statement primarily about equality and racism. But socially and culturally speaking, we could also say, the claim refers back to that earlier idea of democracy which implicitly says we must do away with all forms of negative dominating hierarchy. That idea of democracy *trusts* that we, as a variety of individuals in our various collective groups, belong together—that trust is a necessity for democracy to function.

In democratic societies, there is an interface between the political and the social where trust and belonging reside. The foundational balance between these two factors—social trust and political legitimacy—each reflecting the other—is crucial. When this critical interface is knocked off balance, democracy wanes. In other words, *social trust provides democracy's legitimacy*. Authoritarian states have no such necessity; they depend upon violence and intimidation for enforcement, so there is no need for any such legitimacy. Democratic states, in contrast, because they depend upon the people to collaboratively and socially both accept and enforce existing and legitimated rules and laws, have no necessity for violence and intimidation.

While the 45th former president's loudmouth act, beginning with his loss in 2020, had nothing to do with the long-enduring existence of our inequality practices, his harangues of anger and victimization have much to do with the current fragile state of legitimacy and democracy in our country today. In speaking, his tirades always went straight to the issues of trust and belonging. Trump divided us because his speeches surfaced the true reality of *inequality* in our country and the feelings that accompanied it for many who experienced it. He essentially revealed and manifested that long-accepted but negatively skewed balance of trust and belonging. Who belonged? Who trusted whom?

Trump figuratively held up that *inequality principle* for inspection upon which trust and belonging for white people had originally been based. He effectively reiterated over and over: 'Do you really *belong* in a society that takes away your status by welcoming all?' 'Do you *trust* a government that allows this to happen?' Most of us rejected these questions as nonsense, but those words had an immense effect on a relatively small but now active minority that believed themselves to be increasingly disenfranchised. As a result, in the course of his presidency and after, Trump transformed himself from first, a renegade political upstart that few took seriously, to an initially unwanted hypocritical leader of the Republican Party, and finally, to the *avatar* of the Republican base.

He became an incarnate, almost divine, leader who would give back to those who felt themselves to be disenfranchised the respect they deserved. Because his supporters continued to prefer that *equality* be based on *inequality*, thereby ensuring themselves of respect and dominance, keeping their country as they believe it was intended to be. Once acquiring this status with the Republican base, the political and economic power holders who valued their own status and profits above all else had no choice but to hand over their support to this supposed avatar.

The 45th president's effective out-loud support of that paradoxical principle—"equality depended upon inequality" with its corollary "equality for all will destroy liberty"—is the true reason that his regular outbursts of misogyny, antisemitism, and racism have had such a serious effect on trust and belonging and have divided us so.[15] Because of Trump's words, his base believed their preferred claim to status and belonging in a reinstated hierarchical, primarily white supremacist country would be validated. We now recognize, however, that his base were not the only ones who preferred this status. Those business establishments which house the economic profiteers also prefer this same status—their longstanding King's men hierarchical domination, simply without the drama and trauma of a Trump. But these outcomes are not just isolated damage to a particular political party, economic profiteers, or even the political establishment as a whole.

Once the *interface* of a democratic society's foundational balance between political legitimacy and social trust is questioned and found wanting, it is compromised. Democracy is knocked off-center—and trust, belonging, and legitimacy—all begin to dissipate. Sadly, Trump has accomplished exactly that. He whacked our democracy off-balance and then proceeded to continue its destabilization. Taking the fear and anxiety of some and the necessity for profit and power for others, the 45th president melded a base of people willing to maintain, with violence by some and complicity by others, the ultimate hypocrisy—*equality for all will destroy liberty*. I hate to be ghoulish about this, but if this is allowed to happen, certainly the Excluders will indeed be named as the *destroyers of democracy* by future historians. In fact, Excluders have already declared the destruction of democracy is their intent. However, if the rest of us do not step up to do the right thing, we will become known as the *gravediggers of democracy*! And that is not what any of us want.

Reset for Reality

Choosing to make the acknowledgement of inequality's social acceptance, and beginning to explore how to make things right, means we have faced our critical weakness. We can now begin to strengthen social integrative power and recognize it as part of the foundational interface which keeps democracy alive. The investigation of the four ancient wrongs, the just completed review of their consequences and our present status begins to show us where and how to start. We have begun to see beyond the political/economic manifestations of our difficulties, which so often limits the capacity to understand what's really going on. And we now have better insight into the *social reality* of the situation. As a result, we have decided to change by way of knowledge rather than violence.

The two transforming resets for society identified earlier, "*non-negotiable equality*" and "*social relations as first organizing principle*," move us to a worldview better aligned with the physical reality of the planet, its multiple species, and the true social reality of humans. These two resets will allow all to flourish on a planet that we protect, and which, in turn, protects us. It is these two resets that begin to guarantee justice and dignity for life on earth.

Equality is non-negotiable defines the essential base. No longer can compromises be made if equality is the subject, or difference and indifference invoked, or right-of-command assumed. The message is the same for far-right extremists, moderates, and progressives as well. It is also the same for those who often conveniently tend to be Fence-straddlers, simply thinking of themselves in a positive manner as compromising 'problem-solvers.' If we are to accept "equality is non-negotiable," none of us will any longer be able to hide behind those calls for negotiation, compromise, or move toward the incipient and actual violence that keep exclusion and inequality alive. Of course, at this point, we don't know much about effectively practicing equality. But we can learn.

Some believe the best solution to solve the injustice of inequality is to replace capitalism with socialism. But replacement of one economic system with another economic system still keeps the economy as the *organizing principle* of a nation, culture, and social system. On the other hand, no one really talks about changing societal organizing principles. No, that's not quite true. It turns out that the economist Karl Polyani—still an icon to present-day environmentalists—not only talked about it, he was serious about it. In the 1940s, he actually proposed that economic relations be "embedded" in social relations.[16] The world wasn't quite ready for Polyani's idea at that time, but times have changed.

Social relations as society's first organizing principle requires that the essential base be social. To do so it must recognize that our very humanity is created and conserved through connection and caring reciprocity. This does not mean that economic relations are discarded, only that they are embedded in and follow the necessity for connection and humane action. The practice of equality is clearly not possible through exclusion-based economic relations. Instead, creating a solidarity-based economics will be required.

Luckily, there is a new conversation going on now. Starting around 2010, the increasing numbers of books about justice became a wonderful deluge and included numerous books about US capitalism and its markets. Some of these books focused on how the economic system could be reconfigured to create and maintain greater fairness and equity. Many of the big-name economists and philosophers were on that list—one of them a Nobel Prize winner. They included: Joseph Stiglitz's *People, Power, and Profit*, and his earlier book, *The Price of Inequality*; Robert Reich's *Saving Capitalism*; Thomas Piketty's book, *Capital and Ideology*; Michael Sandel's *What Money Can't Buy: The Moral Limits of Markets*, and his most recent, *The Tyranny of Merit: What's Become of the Common Good?*; and Kopnina and Poldner's *Circular Economy: Challenges and Opportunities of Ethical and Sustainable Business*.

Each one of these books used similar words over and over—"dialogue," "participation," "inclusion," "equity," "equality," and "collaboration." There are even serious and solemn books by economists now attempting to take the 'interpersonal' (read social) into account—but still from the economic perspective.[17] In each instance, the point in using these words is to suggest that certain changes in existing economic systems and practices would enhance fairness and justice. However, from my perspective as a longtime participation practitioner, to even come close to the real participatory practice asked for by the words expressed above—consistently and with practicality—is frankly impossible within our existing highly competitive economic framework that depends upon exclusion. However, none of the economists went so far in their discussions to mention changing the *economic base* to a *social base*, nor have the politicians.

Even so, we are all inching towards this new reality—we just have to do it much faster. Although our society has spent almost the past four hundred years rejecting social alternatives as either too idealistic, not relevant, or not in accordance with nature, things are changing fast. So, while these present-day economists and political practitioners cannot yet step outside our society's longstanding political/economic organizing framework and suggest a new reciprocal social organizing principle for the US society, we can. That's not their job, but it is our job. In a democratic society, it is us, "the people," who must decide on new organizing principles.

Notes

1 Achille Mbembe, trans. Laurent Dubois, *Critique of Black Reason* (Durham: Duke University Press, 2017), 181.
2 A.R. Radcliffe-Brown, *A Natural Science of Society* (Chicago: Free Press, 1957), 87.
3 William Blake, *The Marriage of Heaven and Hell* (New York: Dover Publications, 1994).
4 *Mis-information* is the spread of false information unknowingly. *Dis-information* is the knowing spread of false information to intentionally deceive, manipulate, and mislead. The American Psychological Association has multiple helpful articles on this topic online: apa.org/topics/journalism/
5 John R. Vile, "Stop W.O.K.E. Act (Florida)", *Free Speech Center: First American Encyclopedia*, Nov. 2022, https://firstamendment.mtsu.edu

6 Mark Drozdowski, "Do College Rankings Matter", *Best Colleges*, https://bestcolleges.com
7 Heather McGhee, *The Sum of Us: What Racism Costs Everyone* (New York: One World, 2021).
8 George F. Will, "When 'equity' means hiding achievement", *Washington Post*, Jan. 15, 2023.
9 Bill McKibben, *Falter: Has the Human Game Begun to Play itself Out* (New York: Holt, 2019), 91
10 Dana Milbank, "As Trump is arraigned, Republicans honor the insurrectionists", *Washington Post*, June 18, 2023, https://washingtonpost.com
11 Anand Giridharadas, *Winners Take All* (New York: Vintage Books, 2018), 257.
12 Anand Giridharadas, *Winners Take All*, 257.
13 Anand Giridharadas, *Winners Take All*, chapter 2, 35–59.
14 Herman Daly, *From Uneconomic Growth to a Steady-State Economy* (Massachusetts: Edward Elgar Publishing, Inc, Advances in Ecological Economics), 238.
15 Heather Cox Richarson, *How the South Won the Civil War* (UK: Oxford University Press, 2020), xxix.
16 Karl Polyani, *The Great Transformation: The Political and Economic Origins of Our Time* (Massachusetts: Beacon Press, 2001).
17 Beneditto Gui and Robert Sugden, *Economic and Social Interaction: Accounting for Interpersonal Relations* (UK: Cambridge University Press, 2010).

Bibliography

Blake, William. *The Marriage of Heaven and Hell*. New York: Dover Publications, 1994.
Cox Richarson, Heather. *How the South Won the Civil War*. UK: Oxford University Press, 2020.
Daly, Herman. *From Uneconomic Growth to a Steady-State Economy*. Massachusetts: Edward Elgar Publishing, Inc, 2014.
Drozdowski, Mark. "Do College Rankings Matter", *Best Colleges*, 2022.
Giridharadas, Anand. *Winners Take All*. New York: Vintage Books, 2018.
Gui, Beneditto and Robert Sugden. *Economic and Social Interaction: Accounting for Interpersonal Relations*. UK: Cambridge University Press, 2010.
Kopnina, Helen and Kim Poldner, eds. *Circular Economy: Challenges and Opportunities for Ethical and Sustainable Business*. New York: Routledge, 2021.
Mbembe, Achille. Trans. Laurent Dubois. *Critique of Black Reason*. Durham: Duke University Press, 2017.
McGhee, Heather. *The Sum of Us: What Racism Costs Everyone*. New York: One World, 2021.
McKibben, Bill. *Falter: Has the Human Game Begun to Play itself Out*. New York: Holt, 2019.
Milbank, Dana. "As Trump is arraigned, Republicans honor the insurrectionists", *Washington Post*, June 18, 2023.
Piketty, Thomas. *Capital and Ideology*. Cambridge: Belknap Harvard University Press, 2019.
Polyani, Karl. *The Great Transformation: The Political and Economic Origins of Our Time*. Massachusetts: Beacon Press, 2001.
Radcliffe-Brown, A.R. *A Natural Science of Society*. Chicago: Free Press, 1957.
Reich, Robert B. *Saving Capitalism: For the Many, Not the Few*. New York: Alfred A. Knopf, 2015.
Sandel, Michael. *Justice: What's the Right Thing to Do?* New York: Farrar, Straus and Giroux, 2009.

Sandel, Michael. *What Money Can't Buy: The Moral Limits of Markets*. New York: Farrar, Straus and Giroux, 2012.
Sandel, Michael. *The Tyranny of Merit: What's Become of the Common Good*. New York: Picador Paper, 2021.
Stiglitz, Joseph E. *The Price of Inequality*. New York: W. W. Norton & Co., 2013.
Stiglitz, Joseph E. *People, Power, and Profits: Progressive Capitalism for an Age of Discontent*. New York: W.W. Norton & Co., 2019.
Vile, John R. "Stop W.O.K.E. Act (Florida)", *Free Speech Center: First American Encyclopedia*, Nov. 2022.
Will, George F. "When 'equity' means hiding achievement", *Washington Post*, Jan. 15, 2023.

PART III
Culture Shift Action

12
EXITING THE CIRCLE OF INDIFFERENCE

The question is, how do we become a collective force for equality? Poet-laureate Tracy Smith's description of what poetry accomplishes gives us a clue. She explains it this way: "I'm interested in the way our voices sound when we dip below the decibel of politics … You want a poem to unsettle something. There's a deep and interesting kind of troubling that poems do, which is to say: This is what you think you're certain of, and I'm going to show you how that's not enough. There's something more that might be even more rewarding if you're willing to let go of what you already know."[1]

Once again, we are at that place where true change is possible. And we need to celebrate that. Because of those initial 'stalwart few' that never gave up, we have now become the 'stalwart many' who are on the cusp of possibly creating an equitable and just nation—multi-racial, multi-ethnic, planet devotees, and peace advocates together. There are enough of us to create this new kind of beauty—to collaboratively move forward to full equality and justice. But there is a deeper aspect to all of this which is more complex and needs further exploration. It is, as poet Tracy Smith tells us, those things that, individually and collectively, we have been certain of for so long they no longer require reflection; but if we are honest, some of these may be found to not fit any real practice of equality and justice. So, we need to ask ourselves collectively and individually what exit paths—new ways of thinking, new ways of doing—will truly benefit.

Below the Decibel of Politics

We now understand the intentions of those who practice the four ancient wrongs. But what about our intentions? What are they? How do we negotiate between very different sets of realities and worrying levels of conflict? Are these ideas of fearless change, culture shifts, and social transformation just one big fairytale? Or can deep participation and social power help us do it differently? The answer is 'yes': and there are definitive new ways of putting us back in sync with physical and humane reality. But they rest, to begin with, only within the *perimeter of social power,* not the political or economic. Only when deep participation agreements are reached within these social integrative power boundaries, with their offer of social legitimacy and license to operate, can the same agreements then become operative within political and economic power circles.

Exiting our malfunctioning worldview will require that we explore at least some of what we do and think we know about the social power perimeters in order to find the new future that is needed—transient and moving as it may be. In Chapter 1 introduction, I assured you, the reader, that although I was speaking of the 'social' it did not mean that this book "was just one more riff on love." That remains true. But at this point, I must ask you to focus on the possibilities the phrase "*love ethic*" offers. If there is to be a successful exiting from our current debilitating worldview, this is something that needs to be considered.

Not too many people in our society have the courage to write extensively and consistently about 'love' in non-fiction settings. Being labeled naïve, an idealist, or a lightweight sentimentalist is not what most people want to experience. But Martin Luther King wrote about love: invoking it again and again against the "blind spots" blocking empathy and compassion. He often pointed to a favorite motto—"Freedom and Justice Through Love"—which he taught "can transform opposers into friends." King explained that this type of love meant nothing sentimental or affectionate. Instead, "It means understanding, redeeming goodwill for all ..."[2]

I bring this up because it was bell hook's reflections on MLK and his writings on love that inspired her to define the phrase "love ethic." I like that phrase because it allows us to understand the different possibilities that love offers us—not just as a cherished value or a romantic interlude but also as a strategy for change exiting domination. This is the way bell hooks describes it.

> Without love, our efforts to liberate ourselves and our world community from oppression and exploitation are doomed. As long as we refuse to address fully the place of love in struggles for liberation we will not be able to create a culture of conversion where there is a mass turning away from an ethic of domination ... Without an ethic of love ... we are often seduced, in one way or another, into continued allegiance to systems of domination—imperialism, sexism, racism, classism .... I conclude that many of us are motivated to move against domination solely when we feel our self-interest directly threatened ... The ability to acknowledge blind spots can emerge only as we expand our concern about politics of domination and our capacity to care about the oppression and exploitation of others. A *love ethic* makes this expansion possible.[3]

To understand the true potential of a "love ethic," it is first necessary to remember what has been going on for a long time now. Our current view of how the world works, underwritten by economic and political power, completely *trivializes* anything and everything to do with the social, particularly aspiration, ideals, empathy, love, or compassion. These terms and their expressions supposedly do not, and cannot, belong in the real, pragmatic, dangerous world we live in. If suggested, they are derided.

Characterizations of the social include: unrealistic, baseless, nonsense, lame, and half-baked, among others. Responses to any of those supposedly off-base absurd ideas of social solidarity, etc., are kindly refuted with 'honest straight-talk' emphasized by man-to-man undertones. As a result, these social concepts are off-limits; not to be considered in efforts to solve real-world problems. These aspirations and their resulting actions are to be left at home or in the community meeting room of the church, mosque, synagogue, or temple—anywhere where women undertake their little projects …. Yes, this could turn into a rant, so I'll leave it there. But I am sure you, the reader, get the point.

But it goes beyond gender and sexism—it also engages race, class, and all types of violence. What's really going on here is that at one level the purveyors of conflict and division want to keep us engaged on their terms. A strategic change of emphasis, however, away from these politico-economic divisions, moving to the *social reality* of the situation opens up new avenues of awareness and action.

Changing World Views

Most of us have done our level best to abandon 'negative difference'; but what about 'indifference'? Achille Mbembe forcefully reminds us that even the right to indifference—and that includes the right to ignorance—does not exist. But our society is structured so that the practice of indifference, if not quite necessary, is definitely easy and not at all notable or remarked on in everyday life. Becoming aware of strictures that encourage indifference is a crucial first step.

Certainly, the idea of individual freedom, for example, is essential to both democracy and everyday life. But this freedom, if perceived as only singular and not accompanied by equality, has implicit underpinnings which tend to skew both individual and collective reality. One of them is the vested belief in the insularity and intact nature of the individual. Even the concept of 'individual responsibility' that most of us accept as valid has easy off-ramps which not only allow but also encourage an indifferent shrug of the shoulders with the explanation—'that's their problem, not mine.' Exploring this hyper-individualism and the cultural base which supports this singularity from a historical, as well as near-future perspective, illustrates the captivating, existentially difficult, but beautifully exciting point our society has reached as we now begin to do things differently.

Although we rarely think about it this way, our historical system started with Copernicus, who in 1543 claimed that the Earth circulated the Sun. This one particular claim managed to upset centuries of belief that the Earth was the center of the Universe, and began what is now called the first Scientific Revolution. It continued with Galileo's discovery of Jupiter's moons in 1610. Francis Bacon synthesized and began to popularize these inductive reason findings as a system of logic, starting with his 1620 publication of what came to be known as the scientific method. However, the first broad recognition of the Scientific Revolution began with Isaac Newton's publication of *Principia* (1687), where he established the physics of classical mechanics, which became the scientific gospel of the modern age.

But it was a little-known history professor in 1750, Jacques Turgot of the French Sorbonne, who essentially tied together this new system of knowledge, linking it with the economic and philosophical into a new construct of world history. Fundamentally changing the existing world view, it began the modern era. In his lectures, Turgot very specifically rejected the cyclical theological perspectives of the previous Feudal era as well as the steady-state universe perspectives of the ancient Greeks. Instead, he persuasively—almost combatively— argued that "history proceeds in a straight line and that each succeeding stage of history represents an advance over the preceding one." In other words, history is both 'cumulative' and 'progressive,' with primordial man at its center.[4]

With that, the intellectual "Age of Enlightenment," also known as the "Age of Reason," took off. With further Enlightenment additions, it culminated in a force intertwining economic enterprise and political democracy with a focus on individual rights—for some at least. Turgot brought together the pieces of this puzzle, *framing* a historical tableau within which we all continue to live today. There have been, however, multiple and successive mutations leading to a hyper-individualism that was not originally envisioned.

We are now at another distinct point of change in scientific thought. During the last 50 years of the 20th century, the 1977 Nobel Laureate, Ilya Prigogine, began what many have begun to label the second Scientific Revolution—thus considerably reordering those puzzle pieces. His discoveries concerning dissipative structures and the science of complexity, presented in *Order out of Chaos: Man's New Dialogue with Nature* by Prigogine and Isabelle Stengers, have caused Newton's deterministic universe to be permeated, redefined, and increasingly reordered.

For example, once new theory concepts such as "sensitivity to initial conditions," and "instability" among others, were introduced, confident predictions of the future were no longer possible. Now, the Newtonian "laws of nature," expressed so definitively in classical mechanics, can only be defined as possibilities or probabilities—not predictions. As Prigogine himself observes: "here we go against one of the basic traditions of Western thought, the belief in certainty."[5] He goes on to explain: "In contrast, we believe that we are actually at the beginning of a new scientific era ... reflect[ing] the complexity of the real world, *a science that views us and our creativity as part of a fundamental trend present at all levels of nature.*"[6]

Contrary to the multiple predictions of the 1980s and 1990s, however, this new science of *complexity, chaos,* and *entropy* has not emerged as the 21st century new world view as rapidly as anticipated—it has been instead, slow-going. The Newtonian "Age of the Machine" worldview—with its easy access to total self-interest—is evidently difficult to give up. And we are still waiting for the modern-day Jacques Turgot to arrive and reframe the existing Age of Reason into the Age of Complexity.

Despite the fact that this progression is slow-moving, this new complexity-science foundation is of particular importance to the practice of deep participation and social power because of their strong congruence with this new science. Deep participation's findings and practice are in distinct contradiction to the still current Western individual prototype: 'an insular no-connection, no-relation' solitary entity, accomplishing everything on his, her, or their own. Instead, the science of complexity introduces and verifies patterns of interconnectedness, interacting feedback loops, sensitivity to initial conditions, self-organization, and emergent phenomena—all characteristics of deep participation. All of these together introduce a new and evolving human prototype much more interdependent and inter-relational with the surrounding environment. Deep participation reflects all of these same patterns.

Complexity science, with its multiple feedback loops and connectivity, also specifically signals that the "relational," as well as its correlate "equality," are, for the first time, a requirement in both science and social change. For example, the role that the social factor of equality plays in the science of carbon decrease, as discussed in the previous environmental chapter, is an illustration of this new perspective. Not only requirements but measurements are changing. Calculus in Newtonian science, capturing its multiplicity in a singular measurement, is often described as the straightforward measurement of velocity and change. For complexity science, the conceptual equivalent may be development of a circular measurement that captures the depth and necessity of relatedness, connectivity, interdependence, and equality.

Complexity and the Relational Mind

Particularly good examples of this new 'connectivity' and 'relational' perspectives at work are the increasing number of scientific reports and books on nature and how perspectives are changing. The first Scientific Revolution used images of nature that portrayed competition—think "survival of the fittest." But now Science is learning that cooperation and trust are the greater realities in terms of 'maintaining life.' An example is the role that interdependence plays in the life of trees. Longstanding forest management doctrine focuses on the strength and *singularity* of the tree itself: after a clear-cutting or a fire, neat rows of tiny, new trees are planted, but despite the care given to them, they still struggle to grow. Forest ecologist, Suzanne Simard, noting this, began her own experiments, as recounted in her book, *Finding the Mother Tree*. What she found astounded the, at first, unbelieving forestry establishment:

> Trees share. Fast growing birch send nutrients to slower-moving fir trees. In winter the goods go in reverse. Birch, shorn of their leaves, receive sugars and carbon from evergreens. Mother trees shoot life-giving nutrients in underground mycorrhyzal fungi to saplings circling their crowns. And trees share more than food. They send messages, warnings, and defensive chemicals to neighbors. They form mutual aid societies across species. But they don't collaborate indiscriminately. Mother trees recognize their offspring.[7]

The initial reaction to Simard's research was not at all welcoming. But as book reviewer Kate Brown notes: "The gendered attack on Simard is telling. She was not the first female scientist to make a case for intricacy and cooperation over individuality and competition."[8] These types of negative reactions continue but decrease in number and volume in this exiting phase of the now-faltering Newtonian worldview.

Moving forward with these ideas, the *relational mind* is one of the new and most interesting concepts to aid in our exiting phase. Study of the mind has traditionally always been brain-centric. It is often known as the "brain activity = mind" model, following the Newtonian materialist-reductionist mode of "break-it-down-to-its-smallest-bits." This reductionist mode of thinking is somewhat humorously explained by neuroscientist David Eagleman as the "radio theory" of brains. Imagine that you live in an isolated area of the world:

> You stumble upon a transistor radio … you might pick it up, twiddle the knobs, and suddenly to your surprise, hear voices streaming out of this strange little box. If you're curious and scientifically minded, you might try to understand what is going on. You might pry off the back cover to discover a little nest of wires. Now let's say, you begin a careful scientific study of what causes the voices … You devote your life to developing a science …[9]

In this vignette, 'you' have just opted for the materialist-reductionist mode. 'You' know nothing about far-away distant radio towers or invisible radio waves that carry voices. So, as Eagleman explains, "you would become a radio materialist. You would conclude that somehow the right configuration of wires engenders classical music and intelligent conversation. You would not realize that you're missing an enormous piece of the puzzle."

Of course, Eagleman assures us that he is not saying that the brain is like a radio. But as he points out: it "*could* be true; there is nothing in our current science that rules this out." Eagleman goes on to explain that Newtonian reductionism is clearly not the right viewpoint for everything, including the brain and its relationship to the mind; and to explain this particular relationship, he adopts "emergence": a component of complexity to explain how the whole brain becomes greater than the sum of its parts.[10]

To me, this story captures the exciting exit phase where today we find science, society, and all of us together right now. Seeing the origin of the mind as "relational": not just as a nest of wires encased in a box or a skull; nor embodied as a nest of wires encased in the skin of the body; or even as the "socially constructed mind" long adopted by sociologists, anthropologists, and linguists—all of this changes the basics and begins a new and different transformational process. It goes back to what the poet Tracy Smith enumerated earlier: if we are willing to let go of what we think we know, we may gain something more rewarding.

In his book *Mind*, neuropsychiatrist Daniel Siegel does just that. Neither the "enskulled" nor the "embodied," not even the "social constructionist" brain perspective, as he perceives it, is sufficient to explain the human mind. Using three basic components—*energy, information,* and *flow*—to describe the mind itself, Siegel presents a new working definition:

> Couldn't the mind be considered embedded in our connections with others and the environment, a mind that is not only embodied, but also relational? This stance views mind as both a fully embodied and relationally embedded process. It is not that the brain is simply responding to social signals from others; we are suggesting that the *mind emerges within those connections as well as from the connections within the body itself*. It is these social and neural connections that are both the source and the shaper of energy and information flow.[11]

This perspective is attuned to sociology, anthropology, linguistics, and even Buddhist teachings, but Siegel says it with a physicality that the social sciences or religions cannot. He also indirectly explains something else that is important to keep in mind in terms of how new ideas start to flow but are often blocked. He explains: "modern social neuroscientists appreciate the power of relationships too. Yet even in this division of neurobiology ... the mind is often viewed as brain activity, and the social brain is simply responding to social stimuli ... From the contemporary neuroscience perspective, brain activity still remains the origin of the mind." Siegel is very clear in his departure from this perspective. He clearly states: *the mind is both within and between.*

Siegel recognizes that many of his colleagues still prefer their long-honored 'yes-but' stance. All of this reminds us of the multiple strategies that people have to resist change, those unacknowledged thought-worlds, with yes-but strategies among them. Acknowledging resistance is the first step in not overcoming it, but rather side-stepping it. Most important of all—working within and through the relational, and then leaving that individualized hyper-competition behind, is nothing less than transformational.

Constructing Mutual Genealogies

Accepting and working with the Relational is a crucial aspect of every necessary exit change strategy. The widely shared realization of suffering and injustice, brought about by the murder of George Floyd, has recently profoundly deepened empathy among us, allowing a better understanding of how the relational mind cannot tolerate indifference—and therefore expanding the ranks of the *Includers*. But why now? Why not before, when 17-year-old Laquan McDonald, 12-year-old Tamir Rice, and so many others were gunned down? We all saw it. Once again, I returned to Achille Mbembe for the answer—he always seems to have the right one—a profound response that takes us to that place of empathy, acknowledgement, commitment, and hopefully, the right kind of action.

But first let me explain why Mbembe seems to have so many answers which are both profound and true. Basically, it's the melding of his historical, economic, and social perspectives together that makes his voice so unique. Achille Mbembe's historical perspective begins in Africa with its colonial history; he documents both ancient and recent exploitative histories of injustice. Yet he continues to write compellingly, almost obsessively, about the possibility of creating foundations for a *"mutual genealogy" of belonging*—and that perspective engages me, and I hope you.

In the following quote, Achille Mbembe is explaining the experience of suffering and injustice in general. Yet it fits with great specificity for how George Floyd's murder, or the Uvalde school murders, change personal perspectives. "They afforded the experience of being touched and affected by this brutal exposure to the unknown suffering of others as well as a chance to abruptly exit the circle of indifference in which they had once walled themselves off and to answer the call of these innumerable bodies of pain."[12]

This is what is happening now among many—the reality of it all has finally broken through. But for this reality to stay with us, the attitude of "indifference" must be consciously and consistently opposed. Otherwise, engagement with either "mutuality" or the "relational" is not possible. So, creating this "mutual genealogy of belonging" with its baseline of connectedness and generosity isn't something that can be switched on and off at will—either individually or collectively. It is a reorganization of ourselves as individuals and our society as well. It requires consideration, reflection, and, yes, definitely practice.

Happily, this "mutual genealogy of belonging" also requires the inclusion of *play* of imagination too. It combines—as Edouard Glissant creates in both the *Poetics of Relation* and *Treatise on the Whole-World*—"the discipline of analytical thought with a determined refusal to accept the logic of linear sequences as the only productive logic." This combination allows imagination to integrate and infiltrate into everything—from play to the most serious problem-solving endeavors. Glissant, a Caribbean professor of literature and Nobel Prize nominee, gives many useful descriptions of entering into this space. Another is: "In spite of ourselves, a sort of "consciousness of consciousness" opens us up and turns each of us into a disconcerted actor in the poetics of Relation."[13]

Glissant is impossible to summarize and even difficult to discuss. He is at once a scholarly analyst, a novelist, a poet, and something of a magician. But I bring his work into our discussion to underscore and emphasize that *exiting* from the current worldview in order to begin making equality non-negotiable is not an easy ho-hum passage—but it's not necessarily serious all the time either. Particularly, changing the organizing principle of our society from the competitive political/economic to the humane social requires much more than new policies, better accountability, or new laws to meet the real needs of people. In very basic language, it requires an expansion of heart and imagination, best described in Glissant's poetics of Relation.

Glissant's core focus—in straight-laced English—is the value of diversity and multiculturalism (the real ideas and intentions that these words represent, not just the knock-about mockeries that we have created with them). But it is only achieved through the lens of *Relation* (always capitalized) and always invested with a sense of enduring "social energy." He uses four terms: "creolization"; "trace"; "chaos-world"; and "whole-world" as a base. This vocabulary, for the first time, places us in the same space as the science of complexity outlined earlier. But this time the vocabulary used emphasizes imagination, creativity, and play.

"*Creolization*" is expanded from its original meaning of mixing languages and races to signify *any* society in which different groups interact. "*Trace*" is used to counter the necessity for an enveloping system based on possession; instead, it indicates a path left behind by previous cultures, allowing for a creative wandering. *Chaos-world* is the current clash "of so many cultures set ablaze ... sleeping or transforming themselves, slowly or at lightning speed," whose outcome we cannot yet predict. "*Whole-World* is the possibility of seeing physical diversity and positive difference not as division and conflict but as gifts of creativity and imagination."[14]

Glissant re-characterizes the personal identities of whole-world people as moving from the self-sufficient *root identity*—existing on its own and refuting other tendrils—to the *rhizome identity* "that reaches out to others and is constructed in relation to them." It is in this way he instructs us to re-imagine the world through the Relational; by doing so and practicing, the resultant images begin to reorient our thinking:

> This will be my first proposition: where systems and ideologies have failed, and without in any way giving up on the resistance or the fight that you must carry on in your particular place, let us extend the imagination by an infinite bursting forth and an infinite repetition of the themes of hybridism, multilingualism, and creolization.[15]

So, whether it is complexity sciences, the relational mind, the mutual genealogy of belonging, or the Relational Whole-World—they all point to new exit ramps—guiding us towards a collective solidarity that preserves community as equality while offering freedom as the expansion and creativity of the individual mind and heart. It is at this point that the construction of "mutual genealogies" can truly begin. That earlier 'individual freedom man' with responsibility only to himself, enthroned by Newton, Bacon, and others—passing up those equality/creativity exit ramps one by one—will soon come to the end of that straight-through, system-built super-freeway, initially constructed in the age of Enlightenment.

This approaching end is exactly why we are now seeing, in this 2020 decade, as Achille Mbembe tells us to expect "deep mutations" in our present "regimes of truth." Those who wish to stay on that hyper-individualistic super-freeway need to make up stories to assuage the anxiety they feel as they pass all those 'Exit Now' signs. Mbembe also points out what we all know but can't quite understand: "what is true is not necessarily what has happened or what is believed." But he manages to capture this bleak reality with an understanding of the intense and negative vulnerability of those caught up in these anger- and violence-aligned processes. Mbembe encompasses it all with the force of compassion that is part of that "mutual genealogy of belonging," underwritten by the "love ethic."

> This process has a genealogy and a name—the race for separation and *delinking*, a race being run against the backdrop of a simple anxiety of annihilation. Nowadays a good many individuals are beset with dread, afraid of having been invaded and being on the verge of disappearing. Entire peoples labor under the apprehension that the resources for continuing to assume their identities are spent. They maintain that an outside no longer exists such to protect themselves against threat and danger the enclosures

must be multiplied. Wanting not to remember anything any longer, least of all their own crimes and misdeeds, they dream up bad objects that return to haunt them and that they then seek to violently rid themselves of.[16]

Although it is difficult to contemplate, even the perpetrators, including those King's men discussed earlier, have their share of vulnerability. No child is born cruel and calculating with a penchant for violence. That is taught, practiced, and learned before their eyes as children, and they become what they see. So, for those of us attempting to make an exit from 'negative difference' and 'indifference,' it is important to keep in mind not only the stated political reality but also the largely unstated social reality with its complex relational implications.

Practicing Equality and Liberty Together

We now have more than a simple idealistic belief in the necessity of equality. In this chapter, we have supportive science, changing world views, the imagined possibilities of a relational future, and possible mutual genealogies of belonging. So, is now the time to draw that non-negotiable line of equality within the perimeter of social power? Not quite. Declaring equality non-negotiable may be more difficult than anticipated; certainly, it is more than a simple declaration. It still needs further exploration.

Let's start with the easier aspects, things with which we have had some success. Equity is a popular issue at the moment—at least for the liberal side of the political spectrum. So reviewing how we utilize and operationalize *equality* and *equity* modes in everyday life is a good, if utilitarian, starting point. Because of those political and economic 'inequality compromises' discussed earlier, resulting 'Inequality contradictions' are dealt with every single day. Sometimes they are met with conscious thought and decision, and sometimes simply with thought-world immediacy and dismissiveness. But once a decision is made to compensate for a situation involving this injustice, it is either 'equality' or 'equity' which is utilized to compensate for the recognized inequality deficit. In this situation, both methods are primarily concerned with "justice through compensation"—but use different means to achieve that shared, stated objective.

Equality compensations, at least in our society, mean that each person or group is given the same amount of goods, resources, or opportunities to achieve a desired objective. On the other hand, equity compensations recognize that each person or group, because they have different needs and circumstances, will not necessarily require sameness. Instead, equity allocates the exact amount of goods, resources, or opportunities necessary for a just outcome through distribution and compensation. The portrayal of two young baseball fans, one tall enough to look over the fence at the game and the shorter one needing a bench to see over, is often used to illustrate equity adjustments. In situations such as these, equity's flexible measures create equivalency or equality without sameness. There is now a wide variation of equality expansions, usefully documented through the prism of social equity and how it works in particular circumstances.[17] That is a definition of some measurable success.

The working alliance between equality and equity, while fruitful and useful, only skims the surface. In fact, the alliance primarily exists because of the continuing 'inequality compromises' made in all aspects of US public and private life. These inequities clearly require compensation if some semblance of equality and democracy is to be achieved and maintained. But our country and other democratic countries as well have yet to attend to the deeper but very real consequences of the compromises made between *liberty* and *equality*. This is despite the fact that instability and vague threats of incipient violence, now part of our everyday experience, are clearly the consequences of these same compromises.

Etienne Balibar, in his book *Equaliberty*, outlines a helpful starting point to attend to these consequences. He begins by affirming that "liberty and equality" must be regarded as one term—one principle—that cannot be divided. The history of those original political revolutions in the 1700s in both America and Europe, and the reasons they were fought, explains the necessity for this one undivided principle. Each and every battle was about the monarchies' *absolutism* with its negation of freedom and liberty, and the monarchies' *privilege* with its negation of equality. We don't think about it often, but this obstinate and hated reality was the real and creative force behind the radical idea of establishing "liberty and equality" together in a new relation.

As Balibar tells us, this new relationship between liberty and equality was understood for the first time because of the people's *practice* within those revolutions. As a result, liberty *and* equality together was the forthcoming principle that set both the terms and practice of citizenship, defined civil rights, and awarded the security of belonging to the society through citizenship. To put it more specifically, liberty refers to the right for each individual to make their own development choices, while equality refers to the cornerstone of this liberty—the social community which recognizes the individual dignity of each and mutually supports all with reciprocal care. As a result of this revolutionary practice, Balibar tells us that it was then stated universally: "equality is identical to freedom, and vice versa. Each is the exact measure of the other …." If one is suppressed, the other is suppressed.[18] It was then assumed that a new *political union* could achieve all of that.

Balibar quite candidly observes however, that the manifest *contradictions* of this "language of 'freedom and equality,' or the language of 'oppression and injustice'—whichever language one prefers—has fed the discussions and differences of modern politics since its first moment." He then makes a critical and quite unexpected observation. He states that this ongoing equivocation between freedom and equality is no longer solvable by our political institutions. He flatly declares: "they no longer work in this era."[19]

He tells us that there has been partial but not total success—and this poses both difficulty and danger. "There is a kind of difference that cannot be overcome by the [political] institutions of equality"—simply because it remains external to them ...." These are the "repressed contradictions of modern politics ...." These differences, or contradictions, as Balibar defined them, remain external to the "institutions of a political union" because the differences themselves are *social* and not political.[20] And for all of us, this is the clear danger that must be understood: these contradictions cannot be fixed by current methods of political compensation as we are currently attempting to do with our equity adjustments.

The repressed differences represent true differences. According to Balibar, they include the duality of men and women; the infiltration of family and heredity through race and class; and the division, based on intellectual capacity, of the difference between intellectual labor and manual labor.[21] According to Balibar, these are the true *social realities* of difference as they exist today. But as Balibar explains: they are currently dressed up in various aspects of human biology, historical experience, and memory—necessary because human consciousness demands explanation of the world around us. Equally important, this biological and historical dress-up is exactly where many of the negative thought-worlds of this age began.

Balibar goes on to suggest that a "mutation of politics" may be the only answer—he doesn't say how—but he does point us in the right direction. Actually, he—or any of us—can't say 'how' because it is a matter of praxis and practice. In other words, we will have to complete that passage from 'theory to practice' which Balibar clearly illustrates was the reason for the success of the first revolutions in the 1700s. At that time, our ancestors recognized, with great passion, 'equality and liberty' were *one* indivisible principle. It took the brutal and unjust experiences that initiated the revolutions and the blood, betrayals, and belief within the revolutions of experience and practice to truly understand the 'indivisibility' part of that anthem to democracy—"liberty and equality—indivisible." But still, they could only go part way.

Our ancestors took the first step towards this "Equaliberty," and now we must complete it. How can it be done? Should we just try and practice equality more? Or maybe just try and consciously practice equality and liberty together? The answer is 'no' to both of those because—and we have to stop momentarily and think about this—none of us really know, in reality, what equality is. And this is the crucial differentiation between 'practice' and 'praxis.' Practice means we perform a particular activity to stay proficient in it or get better at it. But how can we do that if we don't clearly understand what equality is, and what it entails?

This is where praxis comes in: it defines the gap between theory and ideas of everyday life—meaning the *practice of new ideas*. Practicing the new idea, during those revolutionary times, of a relationship between equality and liberty is what our ancestors did. Their success was that they did establish equality and liberty as one principle—strong enough to begin a new nation. And now, what we have to do is move from the political to the social to explore the true meaning and practice of equality and liberty—equaliberty.

As I noted earlier, while Balibar does not formulate a solution, he does point us in the right direction. And it is exactly that *"passage from theory to practice"* first formulating the truth of liberty and freedom as one principle, which gives us our direction. Now, to continue preservation of that initial principle of equaliberty we must undertake a second passage that moves from theory to praxis to practice. This second passage will allow us to understand *equality* in all of its facets—not just the half-hearted political, continually renegotiated reality that has been conveniently accepted in this country until now.

How do we start? We begin by consciously exchanging "negative difference and indifference" for "positive difference and reciprocal care." To do so, we begin to practice the ideas just discussed, within the social perimeter. Ultimately, the legitimacy gained in this designated social perimeter is the only way that these same ideas will necessarily hold sway in the political and economic power structures.

There Are Enough of Us

Connecting social relations and reciprocal care will require new ways of seeing. But they are the essentials to making right our relationship with each other and the planet. Moving away from the well-engrained political/economic principles with their commodifying competition to accepting the *social as primary* and then understanding how to make *equality non-negotiable* is neither easy nor automatic. But realizing the necessity of the Relational in all of its aspects helps us realize the wide-open, beautiful nature of this transition, putting us back in congruence with the physical world.

This Relational wonderfully and beautifully includes: (i) accepting complexity sciences with its view that creativity is a "fundamental trend at all levels of nature"; (ii) understanding in nature as in life "intricacy and cooperation" underwrites "individuality and competition"; (iii) and beginning to see "physical diversity and difference not as division and conflict but as gifts of creativity and imagination." Bundled all together, we truly begin to "see with other eyes."

It may not be that difficult either. All we have to do is look around. As a country, we are the best in the world—yes, I know we have all heard that before ... but sometimes we are. Despite our faults, our spoken ideals of "freedom and equality," "liberty and justice for all" have definitively situated our country as the most likely to make good on honoring democracy's necessities. Or, to use Edouard Glissant's words—we best display the themes of hybridism, multilingualism, and creolization that are harbingers of the second science revolution, and the new world's complexity—the community to come. Some of our ancestors were here and suffered the coming of the rest of us as invaders, not settlers. Some of our ancestors were captured, violently enslaved, and forced to take that middle passage. Some came voluntarily, but some of these were too often characterized as "waste people." Some of us came as immigrants and were treated as second-class non-citizens. Some of us came more recently as refugees too often refused passage or rights.

But we are all here now—with our multiplicity of languages, cultures, creativity, and endurance. That is a great positive. All we have to do is stop, look around, and smile at each other. No, this is not naïve idealism—this is the way of the complex and Relational world that we are necessarily moving toward. So, let us understand that the strategies of violence and the strategies of the love ethic are both clearly available. We have the choice to turn from the negatives to the positives. We can choose that two-point reset, then begin building those desired mutual genealogies of belonging.

Notes

1 Ruth Franklin, "The Poem Cure: Tracy K. Smith Believes Bringing Poetry to the Masses Can be an Antidote to our Toxic Civic Culture", *New York Times Magazine*, Apr. 15, 2018, 42.
2 Martin Luther King, "Facing the Challenge of a New Age: 1956 address To Institute of Non-Violence", in James M. Washington, ed. *The Essential Writings and Speeches of Martin Luther King* (New York: Harper One, 1986), 140.
3 bell hooks, "Love as the Practice of Freedom", *Outlaw Culture: Resisting Representations* (New York: Routledge Classics, 2006), 1.
4 Jeremy Rifkin, *Entropy: A New World View* (New York: Bantam Books, 1981), 15.
5 Ilya Prigogine, *The End of Certainty: Time, Chaos, and the New Laws of Nature* (New York: The Free Press, 1996), 4
6 Ilya Prigogine, *End of Certainty*, 7.
7 Kate Brown, "A Career in Communion with Trees", *Book World, Washington Post*, May 23, 2021.
    Suzanne Simard, *Finding the Mother Tree: Discovering the Wisdom of the Forest* (New York: Knopf, 2021).
8 Kate Brown, "Communion with Trees", *Washington Post*.
9 David Eagleman, *Incognito: The Secret Lives of the Brain* (New York: Vintage Books, 2012), 221–222.
10 David Eagleman, *Incognito*, 222.
11 Daniel J Siegel, *Mind: A Journey to the Heart of Being Human* (New York: W.W. Norton, 2017), 47.

12 Achille Mbembe, *Necro-Politics* (Durham: Duke University Press, 2019), 4.
13 Edouard Glissant, trans. Betsy Wing, *Poetics of Relation* (Minneapolis: University of Michigan Press, 1997), "Translator's Introduction, xii, 27.
14 Edouard Glissant, *Poetics of Relation*, 1997.
15 Edouard Glissant, *Poetics of Relation*, 9.
16 Mbembe, *Necro-Politics*, 2.
17 Bernadette Mc Sherry, 2013, "Social Equity", Melbourne Social Institute, online. Some of the areas discussed include: philosophers who have expanded notions of "distributive justice" legal theorists who have attempted to create more "procedural fairness"; public administrators who recognize that administration is not value neutral; and the equality/equity movement continues through numerous other venues.
18 Etienne Balibar, trans. James Ingram. *Equaliberty: Political Essays* (Durham: Duke University Press, 2014), 48.
19 Etienne Balibar, *Equaliberty*, 58.
20 Etienne Balibar, *Equaliberty*, 57–58.
21 Etienne Balibar, *Equaliberty*, 57.

Bibliography

Balibar, Etienne. trans. James Ingram. *Equaliberty: Political Essays*, Durham: Duke University Press, 2014.
Brown, Kate. "A Career in Communion with Trees", *Book World, Washington Post*, May 23, 2021.
Eagleman, David. *Incognito: The Secret Lives of the Brain*. New York: Vintage Books, 2012.
Franklin, Ruth. "The Poem Cure: Tracy K. Smith Believes Bringing Poetry to the Masses can be an Antidote to our Toxic Civic Culture", *New York Times Magazine*, Apr. 15, 2018.
Glissant, Edouard. trans. Betsy Wing. *Poetics of Relation*. Minneapolis: University of Michigan Press, 1997.
hooks, bell. "Love as the Practice of Freedom". *Outlaw Culture: Resisting Representations*. New York: Routledge Classics, 2006.
King, Martin Luther. "Facing the Challenge of a New Age: 1956 Address To Institute of Non-Violence", in James M. Washington, ed. *The Essential Writings and Speeches of Martin Luther King*. New York: Harper One, 1986.
Mbembe, Achille. *Necro-Politics*. Durham: Duke University Press, 2019.
Mc Sherry, Bernadette. "Social Equity", *Melbourne Social Institute*, 2013. https://socialequity.unimelb.edu.au.
Prigogine, Ilya. *The End of Certainty: Time, Chaos, and the New Laws of Nature*. New York: The Free Press, 1996.
Rifkin, Jeremy. *Entropy: A New World View*. New York: Bantam Books, 1981.
Siegel, Daniel J. *Mind: A Journey to the Heart of Being Human*. New York: W.W. Norton, 2017.
Simard, Suzanne. *Finding the Mother Tree: Discovering the Wisdom of the Forest*. New York: Knopf, 2021.

13
DEEP PARTICIPATION PRACTICE

The idea that *legitimacy* and the *license to operate* in a society or culture can only be acquired through social integrative power is clearly a radical idea. Also, the idea that the *deep participation* methodology, when used to bring forth social power, can actually create culture shifts and social transformation will be regarded by some as far-reaching. As a result, this social power concept, always associated with trust, belonging, respect, love, and status, is often ignored—and therefore hidden. While skepticism is necessarily acknowledged, the idea that deep participation and social power can accomplish exactly these achievements is exactly true.

Deep participation is not just one more good idea to give us hope and make a little progress. The six elements of deep participation, fused together, create a dynamic which outlines a new social convention which can then evolve into a new social legitimacy. It has its immediate political advantages also. A major reason often given for the lack of positive change is that our political leaders lack the necessary will. However, leaders of a democracy cannot always be expected to have that necessary political will if the ideas and concepts that formulate the suggested change have not been given social legitimacy by the people themselves. Deep participation is the method of forming that necessary social legitimacy, banishing the negative thoughtworlds and burnishing the positive.

However, I have to be honest—there are no experts here. If you were expecting this chapter to elaborate the six deep participation action-steps for effective social change, you as a sociology student, activist, or observer will be disappointed. While the deep participation methodology itself is new and not well-known, this is not the only reason that I cannot provide these action-steps. In fact, I do enumerate some of that in the final chapter.

DOI: 10.4324/9781003489771-16

But here, it is important that *you,* as a wanna-be Includer, or a reforming Fence-straddler, understand the true depth of social change that is being proposed here. This is important so that each one of you can make your own contributions, growing deep participation into a large and vital deep participation social movement.

So, yes, this is a social movement to join, but it is also a movement that all of you together create—me, I'm just the scribe. No, that sounds good, but it's not quite honest. I didn't just spend the past six years writing this book because I wanted to write a book. No, I wanted to start a movement—one that I've seen work before at the community and regional level; now I want to see it work again, but this time from community to community, across the entire nation.

Currently, as we have explored, our country and society are rapidly moving from long-enduring stability towards full-fledged instability. That is one reason beginning this new and different social movement now makes sense. But that is also the reason there is no Deep Participation step-by-step Action Plan which I can provide—we are no longer in a predictably stable culture, so prescriptions no longer work. There is also another complicating factor. Deep participation methodologies are necessarily culture-bound and people-based which means that they will evolve somewhat differently in different places, different times or around different issues. So, what I can provide at this point in time is a *description*, not a prescription, of the six deep participation elements as component parts of the whole, and how they might interact with each other as you and your group initiate reflection and action. Finally, I can and do specify, with some reliable level of expertise, the critical background components and how they work. Each one of the following sections explore these in some depth.

Retrieving the Good and the True

As a country we have our values, freedom and equality among the most notable, and we are lucky to have them. But even here—particularly here—we do not have clear and sacrosanct agreements on how these, our most cherished values, ethically operate in either our daily lives or in our national and international efforts and actions. In other words, there is a missing *moral legitimacy* that we have yet to collectively agree on. And that missing moral legitimacy is at the heart of our country's divisions today. Deep participation can, however, restore this missing element.

As human beings, we have a deep sense, sometimes somewhat hidden, of the good and the true—that is part of what makes us human. And yet, we do not always agree on what it is. Critical divisions persist because our democratic society has not yet created—built into its social, political, and economic structures—a clear and sacrosanct agreement of what we believe to constitute a 'collective human morality' by which we can all live. I'm not talking about religious morality. There are, of course, numerous versions of religion, and no one version can work for a democratic country; those various versions must remain the choice of the individual.

As a result, we are left with an absence of justice. This everyday lack of collective human justice has created a deep-down ache that many of us did our best to ignore, and some still do. But we comprehend that failure now, and acknowledge our too-long social acceptance of societal inequality and its punishing hierarchies. However, we still have to ask ourselves why we have gone so long without developing a method—a collaborative methodology—to collectively establish the necessary sacrosanct agreement encompassing freedom and equality. There is a quote used by Peter Singer in his book, *Ethics in the Real World*, that begins to clarify this debilitating situation. John Maynard Keynes famously said: "I would rather be vaguely right than precisely wrong." He pointed out that when ideas first come into the world, "they are likely to be wooly, and in need of more work to define them sharply."[1]

That "wooliness" was certainly the case in the late 1600s for science and fact before its clarifying definition. And if we think about it, it is still the case for the ideas of 'freedom' and 'equality' in these early decades of the 21st century, even though they have been in our cultural and political lexicon for 400 years. At the same time, these difficult times are beginning to clarify the "wooly" downsides surrounding our current democratic ideals and ideas by pushing us to rethink and re-evaluate. To assist, John Ralston Saul provides a relevant look back at how that 'missing moral legitimacy,' a major part of the wooly downsides problem, actually happened.

In *Voltaire's Bastards: The Dictatorship of Reason in the West*, Saul describes the late 1700s as the Enlightenment period in full swing. The previously mentioned alt-right King's men and the so-called rabble of the Left were in the midst of forming that 'centrist governance strategy.' At the same time, France's philosophers were busy formulating a new social and political philosophy to escape from the medieval tyranny of the royal monarchy and the Church.[2] These political philosophers—Voltaire among them—joining with their compatriots of Europe, England, and North America, were particularly enamored with *reason*. They became so because they saw reason and science finally breaking "the captive logic of arbitrary power and superstition." Voltaire, for example, wrote in his *Philosophical Dictionary*: "It is obvious to the whole world that a service is better than an injury, that gentleness is preferable to anger. It only remains, therefore, to use our reason to discern the shades of goodness and badness."[3]

That was a mistake that still lives with us today. *Voltaire and his compatriots made the critical error of thinking that reason and morality were one and the same*. In other words, they assumed reason would provide that missing "moral legitimacy." At that time, having clarified how science fits into the world during the 17th century, reason was also regarded as a moral logic rather than simply a disinterested method which clarified science. More important, these philosophers also believed that the application of *reason* to all of the real world—as begun one hundred years earlier by Copernicus and Bacon—was the mechanism that would finally break the "captive logic" of arbitrary power entrenched in the then presiding world of monarchy and church.

They were right in one sense. The overthrow of medieval and monarchial arbitrary power—long supported by the belief in divine revelation and superstition—created a clear and compelling alternative. Beginning in the 16th century this new scientific science alternative was explained and organized through reason. Presenting itself as the 'discovery' option of the physical world, it made it possible to say with truth 'No' to the colloquial misunderstandings of the time—successfully challenging one fallacy after another. Thus, individual mental models and culture shifts began to rapidly change societal worldviews. It is interesting to note it was during this period that social movements first presented themselves—as religious processions—to pressure the ruling classes to move away from their preferred and prevailing status quo.[4]

One hundred years later, as John Saul explains, the Enlightenment philosophers were also partially correct in their embrace of reason. The focus on science and reason had begun to open an extraordinary era of knowledge exploration. But these same philosophers also added *political democracy* to reason's embrace—assuming it to be the final cudgel against the arbitrary power of the monarchy they so avidly sought to destroy. This particular Enlightenment assumption—that the morality of justice and equality is automatically included as part of science and reason—while clearly mistaken, continues today.

That new exercise of secular and democratic power therefore began operating together with little moderating influence of any ethical structure. But evidently no one noticed. At the same time religion, because of its previous hegemony, had been consciously put to one side. So, without a basic human structure of morality to which all agreed or at least were compelled to agree, reason, over time, developed a new subsidiary for morality's replacement. This brand of reason—*rationality*—when applied to science, successfully emphasized the role of neutrality. But this narrow sense of reason was also applied to both the economic and political structures of the new democracies.

As a result, 'rationality' initiated two new orientations when applied to political democracy. The first evolved into a skepticism and cynicism about any belief not directly related to reason and rationality. Ralston Saul explains.

> This skepticism became a trademark of the new elites and gradually turned to cynicism. Worst of all, as the complexities of the new systems increased, with their rewards of abstract power for those who succeeded, the new elites began to develop a contempt for the citizen. The citizen became someone to whom the elite referred as if to a separate species. "He wouldn't understand this." "She needn't know about that."[5]

Deep Participation Practice   **237**

This political cynicism that Saul describes, continues. It has been on display, for example, in both the 2020 and 2024 presidential campaigns. Various media outlets now discuss various issues and present them through two alternative opinion groupings. One category is what they term the 'political elites' with their orientation to policy discussions, presented as of some importance; the second category is who they term the 'regular voters' described as having an affinity for kitchen table issues, often presented with a mild whiff of contempt.

The second rationality mistake concerning political democracy that has evolved is the minimal room for exploration of knowledge outside of what can be quantifiably measured. This has left only limited room for consideration of ideals and the more equitable policies that are dependent upon these ideals. In fact, it has become increasingly clear if we think about it that rationality's continuing utilization has left only a veneer of morality in the corridors of power.[6] A good way to understand the implications of this change is to simply consider definitions. *Reason* is defined as "the process of drawing logical inferences, judgment exercised, and argument pursued." *Rationality* is defined as "the conformity of one's belief with one's reasons to believe."

In other words, reason limits knowledge to what logically can be agreed to, seen, and measured. But rationality goes further; it also presumes *conformity of belief* to these same logical inferences of reason. That would be fine if rationality did not implicitly deny further knowledge exploration because of its preferred methodology. However, it does exactly that, confining ideas and reason itself to the narrow stratum of what can be seen and measured. In effect, it is similar to one of those turn-about arguments based on subconscious thought-worlds that Excluders use in convincing Fence-straddlers to remain on the Excluder side of the fence; i.e., 'ideals are not measurable and therefore not useful in the real world.' Thus, overtaken by narrow bands of rationality, knowledge and ethics are both often hi-jacked.

Consider, for example, the United Nations. It was created in 1945 by 51 founding members, and today has a total of 193 members and two observers, which now accounts for all of the recognized countries in the world. The UN Preamble, in its first two Articles, contains the majority of ideals that humans hold to be most important: "safety from the scourge of war," "peace and security," "fundamental human rights," "justice," "respect," "promotion of social progress," and the list goes on. It is a great organization that has undertaken extraordinary tasks with much success.

It actually personifies an institution based on the Reset principles—"equality is non-negotiable" and "social relations underpinning human dignity as the organizing principle." But these laudable principles and purposes have been, from the beginning, weakened and only minimally attainable. Why? Although the ideals and the words express the world's intent, the *operative* political and economic administrative power structures, similar to all present democratic institutions, do not have the necessary ethical structure. The UN, whether it's the General Assembly with its peace concerns; UNICEF with its focus on women and children; or UNEP with its mission to save the environment, are all good, effective agencies as far as they are allowed. But they also all rely on a bureaucracy limited by rationalism with its restrictive bands of interest in efficiency and neutrality.

In retrospect, science, using reason as its essential 'disinterested administrative method,' has given us 'truth.' That is a critical positive. But it left, 'the good' essentially unattended—to our great impoverishment as a society. Certainly, that "disinterested administrative method' of reason, necessarily accompanying science, has created a society of fact and wealth for us. However, that same methodology of reason, when translated to rationality with its neutrality corollaries of skepticism and cynicism, was, and is, unable to provide a shared ethical structure within either our political or economic power structures.

Because this unattended "good" was rarely missed by any of us, we now have a different, somewhat skewed, society than was originally democratically intended. Similarly, we have a UN organization, as well as so many other democratic institutions, unable to truly follow through on their stated ideals, principles, and purposes because of their rationalized operating systems. So, in addition to our society of fact and wealth, we also have one that features brutal injustices and condescending inequalities despite its stated aspirations and aspiring institutions.

And now we begin to understand. This is why the demand for retrieval of the true and the good resonates with such strength. That is why we march; that is why we protest. Although we are aware that protests remain insufficient—we don't quit—even though we are often unaware of exactly what is needed. But most of all, we don't quit because our country and our democracy have only been 'vaguely right' for several centuries. However, now we know that stance is no longer sufficient. As our country's collective powers to affect and effect the planet and its peoples exponentially expand, we need to be more 'precisely right.' Understanding this background with its more profound objectives faces us in the right direction.

From Practice to Theory[i]

Now once again, conscious effort on our part bears new fruit. Four hundred years ago, *reason* provided the neutral, measuring methodology for the *truth* of science to be identified and utilized. Today, *deep participation,* as a relational and connecting methodology, allows the people to define new legitimacies

---

i This chapter uses parts of the chapter entitled "Six Elements of Deep Participation" from my first book, *Social Justice and Deep Participation: Theory and Practice for the 21st Century,* that was published by Palgrave Macmillan in 2015.

specifying the *good* that a society collectively believes in and wants to act on. That is what community after community under duress in this era has illustrated. Furthering those lessons and learning together how the truth and the good, practiced in freedom and equality, can be inextricably placed within the social perimeter is essential. It is these outcomes, engendering trust, which will enable the building of a true prospering democratic society.

How do we do that? Two research initiatives from two diverse societies have recently identified this deep participation methodology and begins to give us some answers. But this is not a straight-forward story because neither of these research initiatives were attempting to identify a particular methodology, nor were they attempting to question or change social theory. They were simply attempting to identify community initiatives and successes that made things work better within the community and society. In one case, situated in Burkina Faso, West Africa, communities in different geographical areas, as assessed by a Burkina National Team and the World Bank, successfully decreased poverty by relying on, and using their local level institutions in new and different ways. In another case in the western United States, as assessed by the University of California, Davis, numerous local water committees improved community water sustainability, often under difficult conditions.

These diverse communities succeeded in their endeavors because each tapped into, somewhat unexpectedly and unknowingly, the dynamics of what is now recognized as deep participation, and used these processes and dynamics within their local level organizations and institutions. This enabled both the Burkina Faso and U.S. communities to begin transition from that every-day "readjustment change" to the more complex "change-of-type" process judged to be necessary by their communities. Collectively stepping into deep participation dynamics, with its more profound relational and connecting methodology, allowed each set of community groups to identify the transformative change needed and create the collective social energy to attain more difficult and profound community changes. It is this deep participation dynamic that creates social integrative power.

A later identification of this new deep participation methodology these two very different societies both shared, has now begun to initiate differing perceptions concerning of social change and social theory at two levels. First, society's systems and institutions are now recognized to be quite different in times of stability vs. instability. Second, assumptions of societal and community change, long presumed to be based primarily on the well-known and popular conflict tradition, are now found to be questionable at best. Instead, processes used by these communities themselves illustrate that a variation of the Solidarity tradition, with its access to social power and trust, initiated the successful change-of-type action process.

Most research begins with the intent to prove a particular theory, but in this case both of these large well-funded research initiatives began with the preliminary intent to document community practice. This *practice to theory* orientation revealed several critical questions that had not been previously asked or considered. The first, 'why have social scientists focused almost exclusively on the economic and political factors of social change, largely ignoring social integrative power?' The second question is 'why did participation practitioners and sociologists/social anthropologists themselves not identify deep participation?' The answer is the same for both. The preference for one type of research methodology and not another makes all the difference. I realize that this subject is normally not a point of high interest for most of us. But the choice made between "grounded" or "real-world theory" and academia-generated theory is important.

Traditionally, the great majority of research initiatives that have major policy impact have always been quantitative and academically oriented. This is because they are based on a randomization design that emphasizes neutrality and the capacity to generate results which are, in turn, used to define universal generalizations. Grounded, or qualitative research, on the other hand, is primarily based on observation and collection of real-world data, often featuring a particular place and context. For that reason, it has been defined as 'anecdotal,' and therefore viewed as less important.

There is, however, an important distinction often overlooked in this quantitative research process. To successfully design a randomized experiment demands "conscious manipulation by the researcher." In other words, a researcher begins by using current academic theory to define a particular research question to be asked in a specifically designed randomized experiment. As a result, despite best intentions, the outside-academic reality does not necessarily capture what's really going on in a particular geographical area or community.

In other words, the facts and truth of the situation in question are not necessarily well-defined. Other factors, such as the role of all aspects of social power, considered less important or measurable by the quantitative researcher will be excluded; and even if social power is identified qualitatively, it is usually dismissed as of lesser importance. (I must tell you, however, that this is the dry, quantitative research way of defining the situation. I cannot begin to describe to you the numerous, and often extremely funny discussions I have had with economists when trying to convince them to do otherwise. But every so often, an economist would say: "hmm that's interesting, yeah, let's do that.") Raewyn Connell sums up the staying power of quantitative research to the exclusion of other methods particularly well. She observes in *Southern Theory*, "special prestige accrues to theory which is so abstracted that its statements seem universally true—the indifference curves of consumption economies, the structural models of Levi-Strauss ...."[7]

Ironically however, a second reason that social power has been so often disregarded is inadvertently due to participation practitioners themselves, of which I count myself one. Most participation practitioners are well-aware of the social in all of its facets—it is actually the context within which they work. But like every group, participation practitioners have their hang-ups, even their taboos if you like. Comparing and generalizing from one study to another, for example, particularly in search of abstract universals, is not exactly frowned upon but at least questioned. And always insisting that every participatory endeavor be *singular*—based upon the philosophy that every program trajectory very much depends upon honoring the uniqueness of that particular culture—is almost an article of faith.

So, no cross-comparative studies for us. The inviolability of culture is everything. Overall, it is a good precept. It offers dignity and respect to each community and saves us, as practitioners and researchers, from the fallacy of reductionism. This situation meant, however, there was often little effort to compare data from one community initiative to the next. In particular, no one thought to compare the *dynamics* of participation between and among communities. Although most of us as participation practitioners did notice differences, these were most often simply attributed to intensification of community interest. So, although differentiations in participatory power was recognized, comparisons among and between communities and data groups—in other words, comparing the dynamics of participation and looking for differences—never took place.

Despite this lack of interest in comparative analysis, the ongoing and unswerving focus on conserving the "practice first" of each community, a hallmark of all good qualitative, participatory initiatives, was conserved in both the Burkina Faso poverty research, and the California/Washington water committee research. It was this qualitative, practice-first approach, and its focused documentation on *what* communities did, *how* they did it, and *why,* that later allowed the critical discernment of a community's success, in comparison to the participatory dynamics used; and then later allowing further cross-comparison with others. In other words, it was only the qualitative grounded theory approach that documented how communities solved problems, allowing this critical discernment of difference in dynamics and outcome to emerge. As a result, it is now possible to see how the sequence of grounded theory, focusing first on *practice,* and only then moving to *theory* effectively discovers real-world realities.[8]

Recently, this category of grounded or real-world theory and its qualitative results have, for the first time been given high accolades by the 2021 Economics Nobel Committee. The Committee did so for research illustrating that "precise conclusions can be drawn from observation" similar to quantitative-based research. The award went to three economists "for their work drawing conclusions by observing the cause and effect of real-world economic actions." The Committee also concluded that "natural experiments are a rich source of knowledge." Most important, the Nobel Committee stated that this recognition "revolutionized work in the social sciences," essentially backing away from their previous skepticism. Obviously, this is a definitive breakthrough for researchers in general, but particularly for those that rely on qualitative, real-world data. And because that type of data is exactly what deep participation relies upon, new research efforts can, and should multiply.

Real World Starting Points

Background on the two diverse research initiatives involved is helpful. Both of these primary research initiatives were relatively large in scope and had excellent funding. The first research initiative, "Can Local Institutions Reduce Poverty: Rural Decentralization in Burkina Faso," was published by the World Bank in 2001. The West Africa Burkina Faso research, cooperatively initiated by the country's National Decentralization Committee and the World Bank, was supported by funding from the Nordic countries. The research initiative utilized a large national field team under the direction of a Burkinabé experienced in both rural agriculture and the social sciences. He was joined by a two-person World Bank analysis team with sociological, participatory, and economic expertise.

This quantitative and participatory qualitative Local Level Institutions (LLI) study was undertaken in 48 communities and 959 households in four different provinces of the country, with appropriate native language used for each area. A research analysis paper issued by the National Decentralization Committee summarized results. A subsequent joint policy paper published by the World Bank, documents how local communities, using their own indigenous institutions, were able to devise a new social institution and decrease poverty in their communities and surrounding districts.[9]

The Watershed Partnership Project (WPP): Environmental Stewardship in North America was published by the University of California at Davis in 2002. The UC Davis team, with support from the EPA, researched watershed partnerships to assess their effectiveness in Northern California and the state of Washington. These local watershed partnerships, under a variety of names including councils and committees, had been gaining popularity since the federal government began devolving funds to them in the 1980s. This WPP research covered 50 partnerships in the two states. The smallest group interviewed had 6 members and the largest had 76. Membership on the committees was mixed, including local agencies, Native Americans, private sector users, environmentalists, university researchers, and unaffiliated watershed residents, among others.

One of the most arresting and unanticipated findings was that successful rural watershed partnerships, with their diverse memberships, preferred to successfully tackle the more difficult problems in their districts rather than the easy ones. There were other collective successes also. But because the WPP research group itself had come prepared to measure results using more traditional individually-oriented research theories and perspectives, those other collective successes were not fully explored.[10]

The research initiative undertaken in Burkina Faso, West Africa, had unexpected results and impact. The research, initially undertaken with the objective to simply illustrate the multiple and useful roles indigenous institutions played at the local and regional levels, did much more than that. Instead, it illustrated that a new local level institution created at the community and regional level, was key to the successful reduction of poverty in that geographical area.

The first draft of the policy paper on the Burkina Faso research was initially vetted at length by senior World Bank economists. At first these economists were strongly skeptical and questioning of the positive results, particularly those which were partially based on community participatory change dynamics that the LLI national field team had identified. The research and analysis team, disagreeing with this assessment, asked the review economists to join them in a second mutual evaluation of both the qualitative and quantitative data. Over almost a year's work, the initial positive results were verified, partially because the LLI team was further able to verify poverty changes through innovative economic measurements that clearly verified the qualitative results. Because of these findings, the World Bank's Chief Economist decided to present the innovative social and economic research findings at that year's World Bank's Annual Meetings. This signaled by the Bank, for the first time, recognition of the success that a group of local communities had achieved on their own in their fight against poverty. It communicated, also for the first time, an acceptance of social participatory research, with recognition of its effectiveness on par with the highly valued quantitative economic research.

The Watershed Analysis Team of UC/Davis because they ascertained that watershed committee participation processes were shared across the spectrum of committees so they therefore documented it quite extensively. This documentation surprisingly indicated that more profound participatory processes than expected were often being practiced in these rural water committees. Equally surprising, a considerable portion of these same water committees were found to consistently focus their actions on the more difficult water sustainability problems with some success. The UC/Davis Team, through their practice-first orientation, verified that unusual and more profound participatory dynamics were being used across the rural California and Washington state by a number of water committees across both states. But at that point they considered it to be an aberration that could not be explained.

These unexpected results were also inconsistent with the social theory emphasizing 'networking for personal interest' used in the preliminary analysis. Evidently this individual orientation was initially assumed by the UC/Davis Team as the most likely hypothesis to explain differences in water committee success. When that hypothesis clearly did not fit the research findings, no other social hypotheses were pursued. Instead, the results were documented in detail, with the caveat that there was no theoretical explanation.[11] It should be noted, however, that this detailed documentation was, invaluable in the later analyses, allowing the two research initiatives to be comparatively assessed together. So, those real-world starting points are an essential component, even when results turn out differently than anticipated.

OK, if you are a discerning reader—and I assume you are—you are already asking yourself how these two research initiatives ever came together. This is where the story becomes a bit personal, but because it reflects several factors that are important for effective social research, I will tell the story. First, despite the interest of many researchers worldwide, it is still relatively rare for there to be substantial cross-over between US-based and international research carried out on other continents and hemispheres—sometimes we see reports share digitally or, if lucky, share them at an international conference, but that's about it. This is still the status quo today, notwithstanding the environmental, economic, and political global inter-linking of all of us together. Despite this linked reality, as an international researcher, the chances of my being informed or coming across the UC Davis domestic research on my own were close to nil.

However, sometimes the personal helps. Several years after my involvement in the Burkina Faso-World Bank research initiative, my spouse, a hydrologist, attended a US-based West Coast conference focused on rural water committees. Arriving home he asked me: "Remember that work that you did in Burkina?" Saying yes, I laughed; I still thought about it because it remained so intriguing …. Well, he said, "it seems there are rural water committees in California and Washington state that are doing the same participation things that the Burkina groups did." I just stared at him for a moment. No one ever, or at least very rarely, suggests that social processes can be the same between such different cultures. So, we talked, and talked.

As a result, over the next several years, I reviewed the data from both research initiatives and sought out more. I was lucky: because I had worked on so many participatory initiatives for so many different organizations, I had access to a good research base. Analyzed all together, with these two somewhat diverse research initiatives at the center, I was able to define what came to be called "deep participation" and formulate a new theoretical base that differs with several aspects of social theory. This new base calls for use of the Durkheimian solidarity sociological tradition in unstable change-of-type situations, as opposed to the use of the ever-popular Conflict tradition. The book I wrote at that time, *Social Justice and Deep Participation: Theory and Practice for the 21st Century*, brought together the unexpected results of these two research initiatives and their implications for social change.

Observed globally, and operating primarily in the context of rapid social change, this newly identified deep participation methodology illustrates there is a high probability, that the shared, but highly mediated social dynamics, similar across cultures, can introduce successful change in deeply unstable situations. These same hard-to-detect, but very real, participation dynamics are found to promote conscious connection, collective social energy, social integrative power, and new legitimacies. These more profound levels of participation and social power concern themselves with fundamental "change-of-type" situations rather than everyday "readjustment change." Generated by the methodology of deep participation, these dynamics create, through shared critical thought, connection, and collective emotional resonance, definition of new legitimacies that can lead to rapid culture shifts, and slower but enduring social transformations.

Deep Participation's Six Elements

David Bohm, the famous quantitative physicist of the late 20th century, observed that we need to pay "participatory attention" to the world around us. Bohm observes that we don't know how to live together in a changing world; "instead we attempt to live on truths from our past." Accordingly, the resulting incoherence, particularly in our 'modern' world, is endemic. But we do have choices. A colleague further explains Bohm's concept: "Conversely, *collective coherent* ways of thinking and acting only emerge when there is truly a flow of meaning, which starts with allowing many views."[12] Thus, Bohm's words provide a good starting point for becoming a deep participation practitioner.

> The object of a dialogue is not to analyze things, or to win an argument, or to exchange opinions. Rather it is to suspend your opinions and to look at the opinions—to listen to everybody's opinion, to suspend them, and to see what all that means. If we can see what all of our opinions mean, then we are *sharing a common content,* even if we don't agree entirely. It may turn out that the opinions are not really very important—they are all assumptions. And if we can see them all, we may then move more creatively in a different direction. We can just simply share the appreciation of the meanings; and out of this whole thing, truth emerges unannounced—not that we have chosen it.[13]

His point is well-taken. The specific differences among us and between us don't matter as much as we think. Rather, we need to understand how deeply rooted are our socially learned thought-worlds and "existing legitimacies" in both our psyche and society; how incoherent our mutual efforts to communicate them are; and therefore, how necessary it is to establish a more dynamic search for mutual meaning. I liken this decision of beginning a multiple participatory conversation to a martial arts posture. Instead of beginning a fight, we turn aside and let any fist-fighting opponent stumble past—this allows the rest of us to reflect together on *how* we think why we regard some perspectives as "right and true" and others not, and what action need to be taken to guarantee the best of these perspectives.

To arrive at a point where we begin to identify those hidden negative thought-worlds, and then collectively decide what action needs to be taken to either *reform, reinvent, or reimage* our communities and society, we must also move from 'ordinary participation' to 'deep participation.' This is critical; the dialogue of ordinary participation is not enough. Its participatory dynamic is simply a back-and-forth pendulum movement which creates and maintains the everyday 'readjustment change' necessary for the vitality of every group, community, and society that exists within the orb of stability.

But to initiate the more profound 'change-of-type' for those destabilizing communities requires a deep participation dynamic which is very different from ordinary participation dynamics. Instead of that easy, even laconic back and forth movement, deep participation dynamics require group collective thinking processes to confront and solve the newly recognized obstructive realities which appear in destabilizing conditions. To successfully solve such a situation, these deep participation change-of-type dynamics resemble a figure-eight movement, consciously threading up-and over, down-and-through, in order to create a diversity of responses resulting in new social conventions and patterns. Yes, it does take imagination to see the visually unseeable. But with some imaginative effort, deep participation dynamics, with their potential expansion of societal self-knowledge and societal social change, begin to come into focus.

Deep participation does even more. Because its dynamics create a change-of-type process, it is also a basic mechanism for understanding how we think collectively and individually, necessary for the process of social transformation. The six components of deep participation practice effectively create and maintain the balance and resonating emotion between solidarity and critical thinking. Together, they develop shared social learning as the central core and dynamic of deep participation, thus creating a capacity for the previously mentioned *reform, reinvention,* and *re-imaging*. At the same time, altruism and inclusion are identified as central elements of these actions. The determining outcome, *social integrative power*, with its capacity to build trust and confer social legitimacy, appears as the critical product of these elements and interactions infused together. These six elements, presented in short and stylized versions, are as follows:[14]

1 high differentials in economic and political power create destabilizing conditions;
2 social organizations provide the setting for deep participation practice;
3 critical thinking and resonating emotion generate new shared social knowledge;
4 practicing new shared social knowledge begins to develop new social legitimacies;

5 inclusive social energy and collective altruistic action expands this practice;
6 social integrative power confers legitimacy and license to operate as institution.

High Differentials. When culture and power differentials within a culture or society become too high, instability emerges. Every society has these differentials of power; they are maintained by either agreed-to social compacts or, in authoritarian circumstances where there is no social or political agreement, by physical threat-power circumstances. But every society also has a point at which these differentials become destabilizing and detrimental, no matter their method of maintenance. In democracies, tolerance for high economic differentials—exacerbating inequality—may be tolerated, but at the same time, they are always disputed by a minority. Culture differentials, on the other hand—not as easily perceived—are often initially believed by the majority to be a positive in society. In fact, these differentials are like small patches of woven cloth stitched onto the wider swath. For long periods of time, the distinctions are not even noted as the seemingly positive culture façade promotes a *societal unconsciousness* concerning their existence and their meaning. But when tolerances for these unequal differentials in society begin to wane, they're noticed and remarked upon.

When tolerance for either in-your-face economic disparities or the more nuanced culture inequality differentials diminishes, the long-enduring variances imposed by the high differentials will start to be contested. Instability is introduced, and the situation can then rapidly deteriorate. At that point, the social fabric begins to tear around the previously unseen boundaries of difference, and people become aware, start to contest, and begin to separate and divide.[15] Increasing awareness of these community and societal changes begins consideration of what, if any, actions are to be taken.

Social Setting. Social indigenous groups and organizations[16] provide the setting and space for deep participation's generation of new legitimacies. It is only this social placement which specifically and intentionally incorporates the power of *trust* and inclusive *belonging* as the base necessary for participation. This social capacity is in contrast to competitive political and coercive economic organizations and systems. A social placement offering trust and belonging, provides the base for positive individual and collective *connection*. The evolvement and maintenance of "emotion-coated" trust, displayed with culturally appropriate humor and respect, is often a critical signifier. All of this creates a necessarily *inclusive* environment amenable to the tasks of social learning, collective investigation, and collaborative problem-solving. All of these mutual connections make membership trust-creating, personally invigorating, and emotionally sustaining.

Of course, even in social-indigenous institutions, connection and trust may end before they even begin in particular situations. These deviations are usually because of personalities. Some people may decide to flex their power personas and exclude certain ideas or persons; others may attempt to make sure that their perspective dominate at a particular moment. If that happens no matter the original intent—it's no longer deep participation—it's just one more meeting with an agenda. But based on multiple observations of groups over time, social organizations most often do meet the two basic criteria of trust and belonging.

Critical Thinking and Emotional Resonance. The two attributes of 'critical thinking' and 'emotional resonance' fused together provide the capacity for collective social learning resulting in legitimation of new ideas, conventions, and thought-worlds. Thus existing 'negative legitimacies,' operating in nefarious ways, can be successfully challenged and replaced with positive thought-worlds, institutions, and even new social structures. To achieve this, the foundational attributes of critical thinking and emotional resonance can be initiated in culturally different ways but still achieve similar outcomes.

Jurgen Habermas, renowned critical theory scholar, and Paulo Freire, celebrated activist and author, both identify *mutual and participatory discourse as the sole basis* for definition of this critical social reality. They also both agree that "justice, critical thinking and emotion" are the components which create institutional regeneration and legitimacy. Both scholar and activist identify *fusion* of critical thought and emotional resonance as critical, building on the earlier work by Emile Durkheim concerning autonomous and collective social forces.[17]

Habermas indicates that collective agreement is reached only when a mutuality of understanding, shared knowledge, mutual trust and accord with one another, is established. Freire agrees, but he describes it differently. He starts with critical thinking but he adds a nonreligious version of love, humility, and faith. *Love* is an act of courage that commits one person to the next; *humility* is required for the task of learning and taking action together; and intense *faith* in humankind—its ability to become more human and its power to make and remake the world. This element of critical thinking and emotional resonance illustrates, in particular, the necessity for the Solidarity rather than the Conflict tradition as base.[18]

Collective Social Learning. Because creating this fusion is central, deep participation requires an altered understanding of mutual and shared social learning. We already understand that deep participation requires placement within social-indigenous institutions and organizations—in other words, within the social perimeter—thereby guaranteeing the base of trust and inclusion. However, the 'social learning' which emerges from this process is quite different in its intent and usage when compared with individualized learning. Social learning, because it is collectively created, and then mutually accepted as true and good, creates much of that sub-conscious base where thought-worlds, positive and negative, as well as incognito societal structures all reside. Much of this is passed forward to the individual without conscious intervention But new social learning, collectively and currently consciously created, can also be used as a base to define and redefine the shared social, political, and economic life around us—including the organizing principles which shape our world.

In these situations, acceptance of deep participation practice and its shared social learning potential for fusion does definitely depend upon some better understanding of social cognition processes as the foundation of our society—or any society. One thing is clear: the sub-conscious thought-worlds and institutions under discussion here *don't think* for us, as some have claimed. But we do offload a substantial amount of our early cognitive learning, formation and daily life decisions to these same entities. Mary Douglas, one of the most consequential thinkers of modern social anthropology and a creator of thought-world and institutional theory, explains: "Classifications, logical operations, and guiding metaphors are given to the individual by society." Consequently, "the sense of 'a priori rightness' of some ideas and nonsensicality of others are handed out as part of our social environment."

For efficiency's sake, that sense of 'rightness' is perceived as an individualized experience—in our society this is easier to comprehend and believe. The more difficult-to-accept idea is that these same thought-worlds have anything to do with our more collective sense of right and true, working as they do, "out-of-sight, out-of-mind."[19] However, collectively using deep participation methods allows us, as enumerated separately by both Habermas and Freire, to consciously and collectively work within this shared social learning framework to create the necessary fusion leading to desired change.

Social Energy and Altruistic Action. The phrase *social energy* best captures the intangible but active altruistic orientation and action which accompanies all participatory endeavors. Stepping into and experiencing an energy field created by groups of people is part of everyone's everyday experience but we rarely recognize it as such. We might have had, for instance, the experience of being at a dinner table with good friends eating delicious food, shouting out ideas, laughing at past memories, and spinning future dreams. Another time, we may feel a similar energy when we work with a group of people distributing food and water after a climate catastrophe. All of these are short vignettes illustrating that social energy is everywhere we find collective experience, effort, and enjoyment.

However, we rarely, if ever, consciously sit down and cognitively recognize such social energy experiences as something that can be separated from the context within which they are experienced. Particularly in northern cultures, the idea that social energy can be paired with ideas and ideals for collective societal re-organization is rarely understood; primarily because we tend to focus on individual leadership, and discount collective experience. As a result, social energy remains difficult to recognize. But group participation changes that. We learn as we experience. The fusion of critical thought and emotion first elucidates a *shared meaning* within the group. After that, the potential legitimation of this shared meaning requires collective action and practice. It is here that acting together in generosity brings forth a joy, vibrancy, a delight that lightens and transforms the labor.

If we think about it, we can all bring up experiences similar to those mentioned above. But once we become even slightly aware of this collective and essentially joyful social energy it becomes more apparent. As I recount in my earlier book, one of the first times I consciously understood and experienced the true forceful capacity of these social energy fields was in a small village in southwestern Burkina Faso. A group of us, from the capital city of Ouagadougou and the Institute of National Education were meeting with local parents and teachers to discuss new nationally proposed education changes.

After a long day of community meetings, I was sitting to the side, slightly fatigued from listening and interacting in a language I did not know well. But I was also aware that my Burkinabé Institute colleagues had been quite successful in their multiple participatory presentations and discussion with the community. As I sat there, twilight moved in fast, the moon came up, and the discussion tone seemed to soften. As I began to focus once again on the meeting's discussion, everything seemed to slow down and I began to "see" this intangible but very real collective social energy field slowly building itself up around the group of local parents and teachers as each decided to move into the proposed collective community initiatives for the betterment of their local schools.

The memory of that experience stays with me—still with the force of its social energy—and now it's easier to see this same social energy in diverse places: on the dance floor, in singing groups, when building new ideas in meetings, and in the collective rhythms of altruistic ideas coming together. It's always there for us to use, but only if we throw off our societal self-induced sole reliance on the individual. As Kristen Monroe observes, this limitation "is not necessitated by empirical reality."[20]

Social Integrative Power. This sixth and final element of deep participation is part of the practice lexicon—and also its culmination with its capacity to engender trust. Collectively questioning longstanding legitimacies and building new ones requires a transparency and an openness on many different levels to all people, organizations, and institutions. When considering this element of practice, it is helpful to remember Kenneth Boulding's observation. He reminds us the single most important element in determining the long-term viability of any institution in society is its perceived social legitimacy—"no institution, pattern or behavior, or role structure, can exist very long without it."[21]

All of this together begins to rearrange our understanding about the role of social change, legitimacy and how it comes about. In other words, we have to truly comprehend the necessity of collective trust: think it, feel it, and align it with our collective morality of justice if social power is to effectively legitimate new systems, structures, and institutions. The critical components are always basic human trust and morality, and the critical methodology for collective agreement is always deep participation.

New Game, New Rules

But that's not the end of the story. It turns out that the implications of deep participation practice and theory were not well initially understood. For example almost all of us tend to automatically assume a certain level of *stability* to be present in every society, unless they are clearly involved in conflict. Although West Africa certainly had its political and economic challenges to deal with at that time, the cultures themselves seemed to remain quite stable, particularly in Burkina Faso. In California and Washington State there was also certainly no reason to question their stability—again it was automatic.

But for me personally, that perspective began to change as I worked more often in countries even mildly affected by conflict. There I began to understand, particularly with insights generously shared by African colleagues from diverse countries, that instability is there unseen, long before it becomes visible. With this experience, when the US political process in 2015–2016 came around, I viewed it somewhat differently than most. As it unfolded, I slowly became more and more uneasy. Partially, this was because the Republican candidate reminded me of the strong-arm politicians who had emerged out of the long-enduring post-colonial chaos left by the colonizers on the African continent. But equally important, the instabilities of inequality and negative difference of my own country were also increasingly on display. As time went on, I realized that we, as a country, also had our own post-colonial chaos, however distant, and this candidate was taking advantage of these inequality fault lines.

As a result, I began to rethink the implications of deep participation practice and theory in terms of stability, instability, and the divisions leading to conflict. To begin with, I realized the presumption of stability was too often a mistake. Does the societal landscape change when we begin to assume the possibility of *instability*? The answer is, of course, yes. It changes in terms of the perceived problems themselves, ideas on how to solve them; including from the social science perspective, the process of theory definition, the pros and cons of qualitative vs quantitative data, and the use of participatory dynamics.

But most of all, for everyone, the perspectives on trust change. When trust diminishes, instability increases. And I slowly realized if we are not sufficiently aware of this diminishment—over a period of several years or even decades—this instability can be used as a cover for political takeovers. During this 2020 decade, I also began to understand that it's not the takeover leader per-se who is responsible for the lack of trust and increasing instability. Instead, the takeover person is just someone who knows how to take advantage of the existing trust deficits in the society and amplify them. This particular insight turns out to be critical in terms of deciding what future deep participation collective actions will best stop this destabilizing process.

Reviewing the deep participation research from this perspective, three things became clear to me, and I believe they will be valuable to you also as deep participation initiators and participants. First, from both action and theoretical standpoints, not enough emphasis has been given to the differences between *re-adjustment change* which maintains stability, and *change-of-type* which creates substantive change that can alter, sometimes radically, the understood order of how things work. Second, there is often not even enough awareness that there are real and substantive differences between these two types of social change themselves. But awareness of this difference is critical if we are to have any success in creating culture shifts. Third, it is here that the notion of *praxis*—putting theory into practice—comes into clear focus.

Assumptions of stability allow us to use tried and true remedies for change, adopting either political and/or economic power—the differences are only about which one. On the other hand, recognizing instability requires utilization of new and creative remedies for which we have little or no experience. For this, we turn to praxis, the practical application and practice of a theory or an idea. When this begins, shared social energy and altruism build through mutual creativity. But we are lucky. Currently, we have already taken the first step in potentially eradicating the four ancient wrongs by identifying the intertwining problem to be solved—that is socially accepted *inequality*. So, we have a head start for successful praxis of a quality that we easily enumerate—*equality*—but have little true experience in its application and practice. However, as our culture's long-term acceptance and acquiescence becomes better recognized, it becomes easier to attain those culture shifts.

In retrospect, almost everyone has interpreted social change from the context of assumed stability, or at most by only recognizing cycles of mild instability which given a 'nudge' will return to stability. As a result, even deep participation was initially viewed as a mechanism to "fix" failing aspects of relatively stable societies by giving some nudges in the right direction. It can certainly do that. But that is not the primary role of deep participation. In contrast, it is not simply a mechanism to ameliorate negative aspects of stable societies or "fix" the worst components of failing societies and communities. Instead, it is, or at least it can be, a culture-shift mechanism to institutionally and structurally change an unequal, failing democratic society from the inside-out by establishing a new more equal social order. That means it is a new game with new rules, and once again we're all in it together. So, each of us has to decide how they will meet these changing times. For me, I decided to write this book. What will you decide to do? What are you already doing? And most important, what is it that we can all do in the near future?

Courage in Solidarity

Unequal societies often manage to remain sufficiently stable despite their inequalities because they place guard-rails around sub-conscious thought-worlds, guiding institutions, and ideas that would upset stability. They do so by posting no-trespassing signs: "no-interest," "not relevant," or "not possible" types of disapproval signs. This does not mean these societies have identified the ideal and are satisfied. They are just very good at circumventing change. In these situations, zones which are recognized as off-limits to the "public" are often placed there by the current faction of the Kings' men, working as always to guard their preferred status quo. But at the same time, we can choose to interpret these signs quite differently. From this more courageous perspective, those "No Trespassing" signs can often indicate the paths to actually take.

It helps that we now know what deep participation is and how to use it. But still, when we are confronted with the above types of full-force stops, we tend to honor them with remarkable levels of acquiescence. So, it helps to remember what Tracy Smith, the poet, told us in the last chapter about being certain—that certainty is not enough. She went on to say: "There's something more that might even be more rewarding if you're willing to let go of what you already know." And that is what deep participation can do for all of us—it encourages us together, to let go of the negatives that we all know so well, so we can begin to better understand the positives which we are just beginning to discern and understand. Remembering this helps us move forward on the exit paths of transition. It gives us the courage to use social integrative power and begin those social transformations that will lead us to the culture shifts we all want and need. That shift which creates a society featuring the interdependence of equality, reciprocal care, and the freedom of individual creativity.

Notes

1 Peter Singer, *Ethics in the Real World* (Princeton: Princeton University Press, 2016), 196–197.
2 John Ralston Saul, *Voltaire's Bastards: The Dictatorship of Reason in the West* (New York: Vintage Books, 1992).
3 Voltaire, *Dictionaire Philosophique*, vol. 6, (Paris: Librarie de Fortic, 1826), 307, in *Voltaire's Bastards*, 31.
4 Charles Tilly and Lesley J. Wood, *Social Movements: 1768—2008* (Boulder: Paradigm Publishers, 2009).
5 John Ralston Saul, *Voltaire's Bastards: The Dictatorship of Reason in the West* (New York: Vintage Books, 1992).
6 Saul, *Voltaire's Bastards*, 31–37.
7 Raewynn Connell, *Southern Theory: The Global Dynamics of Knowledge* (Oxford: Polity Books, 2007).
8 Paula Donnelly Roark, *Social Justice and Deep Participation: Theory and Practice for the 21st Century* (UK: Palgrave Macmillan, 2015).
9 Paula Donnelly Roark, Karim Ouedraogo, Xiao Ye, "Can Local Level Institutions Reduce Poverty?". *Policy Research Working Paper*, 2677, Washington D.C.: World Bank, 2001.

10 P Sabatier, J Quinn, N Pelkey W Leach, "Watershed Partnership Project: Environmental Stewardship in North America, (Davis: University of California, 2002).
11 P. Sabatier, et al., "Watershed Partnership Project", 2002.
12 David Bohm, *On Dialogue* (New York: Routledge Classics, 2004), xi.
13 David Bohm, *On Dialogue*, 30.
14 Paula Donnelly Roark, *Social Justice and Deep Participation*, 83–84.
15 Paula Donnelly Roark, *Social Justice and Deep Participation*, 84–86.
16 The phrase "social organizations" often suffice as a descriptor. However, to be inclusive of organizations that hold and recognize their indigenous history in high esteem, the more descriptive phrase "social-indigenous" is used here.
17 Emile Durkheim, trans. Carol Cosman, *The Elementary Forms of Religious Life* (UK: Oxford University Press, 2008), Introduction by Mark S. Cladis, xxi.
18 Jurgen Habermas, trans. C. Lenhardt and S. Weber Nicholsen, *Moral Consciousness and Communicative Action* (Massachusetts: MIT Press, 1995), 200.
    Paulo Freire, *Pedagogy of the Oppressed* (London: Continuum Press, 1997, 2017), 73.
19 Mary Douglas, *How Institutions Think* (New York: Syracuse University Press, 1986), 45.
20 Kristen Monroe, *The Heart of Altruism: Perceptions of a Common Humanity*, (New Jersey, Princeton University Press, 1996) 236.
21 Kenneth Boulding, *The Economy of Love and Fear* (Belmont, CA: Wadsworth, 1973), 3.

Bibliography

Bohm, David. *On Dialogue*, New York: Routledge Classics, 2004.
Boulding, Keneth. *The Economy of Love and Fear*, Belmont, CA: Wadsworth, 1973.
Connell, Raewynn. *Southern Theory: The Global Dynamics of Knowledge*, Oxford: Polity Books, 2007.
Donnelly Roark, Paula. *Social Justice and Deep Participation: Theory and Practice for the 21st Century*, UK: Palgrave Macmillan, 2015.
Paula Donnelly Roark, Karim Ouedraogo, Xiao Ye, "Can Local Level Institutions Reduce Poverty?". *Policy Research Working Paper, 2677*, Washington D.C.: World Bank, 2001.
Douglas, Mary. *How Institutions Think*. New York: Syracuse University Press, 1986.
Durkheim, Emile. trans. Carol Cosman. *The Elementary Forms of Religious Life*. UK: Oxford University Press, 2008.
Freire, Paulo. *Pedagogy of the Oppressed*. London: Continuum Press, 1997, 2017.
Habermas, Jurgen. trans. C. Lenhardt and S. Weber Nicholsen. *Moral Consciousness and Communicative Action*. Massachusetts: MIT Press, 1995.
Sabatier, P. and J Quinn, N Pelkey W Leach, *Watershed Partnership Project: Environmental Stewardship in North America*. Davis: University of California, Davis, 2002.
Saul, John Ralston. *Voltaire's Bastards: The Dictatorship of Reason in the West*. New York: Vintage Books, 1992.
Singer, Peter. *Ethics in the Real World*. Princeton: Princeton University Press, 2016.
Tilly, Charles and Lesley J. Wood, *Social Movements: 1768—2008*. Boulder: Paradigm Publishers, 2009.
Voltaire, *Dictionaire Philosophique*, vol. 6, Paris: Librarie de Fortic, 1826.

14
ORGANIZING FOR CULTURE SHIFTS

Personally, I like protest marches and go to them with family and friends. Mostly it's just marching along together, with some chanting interspersed, and then standing listening to speakers and clapping. But every so often, it becomes more intense. Several years ago, this happened at a march to support immigration in a small coastal city in northern San Diego County. The turn-out was quite good, but the organizers did not realize that they needed a permit for all of us to gather in the park. So, the police were there, in force, to disperse us.

We gathered around one of the organizers as she announced apologetically that we had to go home. There were a few groans, and someone loudly muttered, "we shouldn't have to do that." I raised my hand, looked around at the crowd, and asked: "Why should we go home? This is a small town; they can't arrest all of us. The jail is too small!" There was silence for a moment, a few gasps, laughter, and then both the police and TV news people began moving toward me. We didn't go home. Instead, we moved out of the park onto the street, where no permit was needed. The next weekend, at a lunch together, my protest friends gave me a T-shirt stamped "Thug Paula." My husband laughed, rolled his eyes, and said: "she'll never learn."

Yes, protests are often quite enjoyable and there are always funny stories that remain in our memories. But there is a key question about every protest march, and for every marcher. That is: what happens next? Our group would often go to a nearby restaurant or a bar after a march and inevitably strike up a conversation with other marchers. When someone asks, "what's next?" the answer is always the same: a slight shrug of the shoulders, perhaps a smile. And sometimes someone would raise their glass and say—"here's to the next march!"

DOI: 10.4324/9781003489771-17

In other words, despite strong and good intentions, few of us have any idea of what to do next. Yes, we may contribute to our favorite political organizations, or we may volunteer at our favorite voluntary organization—there are so many good people doing so many good things. And situations change fast too. In the summer of 2024, a massive jump in political energy was seen and experienced by many when Vice President Kamala Harris entered the presidential race as the new Democratic candidate. So, that changes the *political* calculus in numerous positive ways, and if Vice President Harris does win the presidency, it will change the *social* calculus. But only slightly: Harris will make it easier to create the social power that will take down inequality, and Trump will make it harder, much harder. However, either way that social power to create the needed culture shifts and the reinvigorating social transformations remain with us, the people. And at this point that social power is most effectively packaged in a new kind of social movement.

Traditional Social Movements

This new kind of movement is quite different from traditional social movements.[1] This is because traditional social movements primarily operate with the intention to change critical aspects of the political landscape—more impact on climate change; more rules against police violence; more funding for child poverty—all of which can be defined as evolutionary or adjustment change. This change makes the existing situation better, but it does not change the societal structure and how it operates. Even when the holders of political and economic power acquiesce to social protest demands, it remains primarily on their terms. This is not a new idea or understanding of political reality. Years ago, Piven and Cloward in their famous and now classic book, *Poor People's Movements: Why They Succeed and Why They Fail,* made this discouraging fact quite clear when they concluded: "protesters win, if they win at all, what historical circumstances have already made ready to be conceded." In other words, "what was won must be judged by what was possible."[2]

When organizing to protest a specific wrong, for example, the hope is that it will increase the political will to fix the wrong. It thus energizes and builds momentary solidarity for the people involved. However, that resolve or commitment, mainly created from shared defiance, tends to be short-lived. So once participants leave the group—the doubts creep in. As individuals, no matter our collective defiance and the immediate success of the protest, we begin to doubt our power to really change things, so then acquiescence sets in. It seems as though as soon as one hard-fought wrong is partially righted by massive traditional movements, another similar wrong follows hard on its heels—think George Floyd and the aftermath of continuing police murders of black men. Hard-core activist leaders know it's difficult, if not almost impossible, to keep that momentum going.

Scholars also know this. Traditional social movements and social protests, always designed to effect political and economic change, have been a favorite subject since the 18th century. I have, for instance, a 10-page reading list for college graduate programs which includes 110 book and article titles on social movements, divided into five different subsections. There does seem to be missing data, however. For example, within the above-mentioned reading list, there is only one—only one—book that mentions human morality or ethics as part of the protests. In addition, there are only two articles that look at participation of any sort.[3] Despite this thin representation, moral protest and participation have been the critical substance of social movements since their beginning in the 1600s. People's moral outrage against wrongdoers is perennial, as demonstrated by the long endurance of the more successful justice movements against racism. So, people, despite the obvious difficulties, have continued to march, now in ever greater numbers, and protests continue to multiply.[4]

These multiple types of non-violent protests, despite their shortcomings, have had success, and it's important to understand what they have accomplished. *In Civil Resistance: What Everyone Needs to Know*, Erica Chenoweth analyzed the success of all efforts around the world to overthrow repressive national arrangements between 1900 and 2019. She found that over 50% of *non-violent movements* succeeded, while only 26% of the violent ones did. Chenoweth observes: "That's a staggering figure that undercuts a widespread view that nonviolent action is weak and ineffective." Of even more immediate interest, Chenoweth's research also illustrates if only 3.5% of the population joins a non-violent social movement, success is highly probable *when* the *movement itself* is sustained (emphasis added). In this context, it is interesting to note that research data indicates that 20–30 million people in the United States joined protests against various forms of injustice between 2015 and 2020.[5] This means, applying Chenoweth's same percentages to the US population—(3.5% of the US 330 million population is 11,550,000 people)—that there are already more than enough people involved for a non-violent social movement to be successful.[6]

Do the newer traditional social movements, now referred to as "new social movements," have the ability to expand these non-violent movement-based accomplishments? The answer is sort-of, or yes-but. Going back a bit in time, the most famous of these social movements, known by the name of the city which launched its world meetings, Porto Allegro, Brazil, had more than 150,000 attendees at its high point in 2005. Currently known as the World Social Forum, things have changed. It had only 2,700 attendees at its last annual meeting in 2023. While it still retains some of its popularity, with 46 ministers, 50 ambassadors, and 15 mayors attending, it has also run into major problems.[7] These "new movements," while social in their focus to alleviate injustice, still continue to use political strategies—and only political strategies—to achieve their objectives. This is the problem. While they do recognize that perennial moral outrage against wrongdoers, they do not and cannot adequately incorporate basic human morality with its accompanying social legitimacy into their outcomes as long as they only use political or economic strategies.

Chenoweth's preliminary data also indicates, similar to the World Forum situation, that since 2010 the success of these traditional non-violent social movements may be waning. Why? Part of the answer may be that as we become more aware of the world around us, the continuing "absence of justice" becomes increasingly apparent and sometimes overwhelms. But, while acknowledging that not enough change in terms of justice and equality takes place, still it seems there is nothing to do but march—both here in the United States and across the world. The brave journalists and diverse communication technologies; the global alliances which pursue injustice claims against any one nation for its victims; the realization that justice is real, necessary, and not a futuristic ideal; and finally, the willingness to move beyond the abstract view of a particular legal and economic framework—all of these contribute to the numbers of protesters and protest marches that continue. We can, however, become much more effective.

Deep Participation Social Movements

Deep participation social movements are different. Instead of changing political regimes as some countries must or even the attempts to change specific policies of injustice as we often do in this country, deep participation social movements operate with the intention of first making a clear change-of-direction within the *social* landscape. By doing so, they begin to gain new legitimacies for justice and humanity. And along with these new legitimacies, they gain a critical leverage to re-evaluate those shared, collective thought-worlds to which we unconsciously adhere and which accompany any proposed action. This re-evaluation, collectively and inclusively undertaken, establishes, as best as currently possible, the true and the good. This injects the legitimacy of the social with its capacity to generate *trust* as primary, and places political/economic power as secondary.

Using those previously identified guideposts as starting points: (i) "equality is non-negotiable" and (ii) "social relations, not economic relations, are society's primary organizing principle," make good shared guides. But one point is particularly important here. It's fairly easy to understand the phrase "equality is non-negotiable" because we see so many efforts to achieve that goal. But the idea of changing our society's organizing principle from the economic to the social is more difficult to grasp.

So, let me give one more innocuous example of the kind we hear every day. This situation happened to be on the front page of my daily newspaper this week. The head line reads: "The Concussion Files: Broken promises, Shattered dreams." The article reports that despite a National Football League legal "concussion settlement" with their players' union, hundreds of players, black and white, now suffering, and some dying, from related dementia, have been denied payments for the care they were promised by the contract with the NFL.

The reason that the NFL has been able to get away with this is because our society is primarily organized on economic principles. In other words, it is the economic power to win that is honored and what counts—not justice nor recognition of social connection or any responsibility for reciprocal care. Yes, the corporate NFL honchos realize that in this economically organized society, they may be sued. But they also know they have the money *and* the time to withstand these pushbacks. But most of all, the NFL corporate heads count on the fact that most of their audience will view this situation with "indifference," if not "negative difference," blaming and minimizing player requests and calling into question the legitimacy of their complaints.

So, yes, all of those same negative thought-worlds are once again doing the dirty work of the King's men. And once again they use those double-talking phrases of 'socialist' or 'communist' to bring up all those still existing negative thought-worlds in our collectively shared thought-worlds. And once again if we stop and think about it, we are reminded that those economic principles around which the King's men organize themselves deliberately limit terms of membership solely for profit. No matter that many of these football players were once held in high regard, some even as heroes. But now no more; they are no longer economically useful. So, who's next?

On the other hand, if our country was primarily organized around the primacy of social relations, audiences would immediately call the NFL out, proclaim injustice, and boycott the next National Football League games. As a result, the NFL would quickly rethink their actions. But this current situation exemplifies what happens when a society is organized around economic relations rather than social relations. Economic relations always support not what is the just or right thing to do from a human perspective but only who can win in a no-holds bar economic power competition over any semblance of justice every time. And this happens every day in multiple guises.

But we are so accustomed to it, we don't even notice the accepted *economic structure* holding it in place. Finally, I will note the headlines on either side of this NFL/Player story without comment. You, as reader, can make your own assessments. To the left of the center story: "Railroad industry slows down on safety"; and to the right "Dozens killed in U.S. strikes, but little damage to Tehran's assets."[8]

Alternatively, a deep participation social movement—defining truth by 'fact and critical thinking,' together with the good of 'connection and conscience'—creates both justice and a missing moral legitimacy. It is only when these two factors are brought together that a particular concept has collective positive standing. Only then can it begin to permeate the structure and system of the society. It is this combination of the reasoned 'truth' with the relational 'good' that gives deep participation social movements the leverage necessary for real change, which we then name culture shifts.

The Enlightenment philosophers spoke the words of freedom and equality, as did the founders of what is now the United States of America. They were both "vaguely right" in an exhilarating new fashion. But they were only the *words* of ideals and reason 200 years ago and so they remain the abstractions of truth upon which democracy is built today. This has made for a "cerebral patriotism" now clearly under attack by those that prefer the more cruel affinities of the Excluders, expressed as tribe and clan—with all of their negatives of exclusion and conflict—putting democracy in danger.[9]

So, it is here that we use deep participation social movements to intervene. By joining the abstract ideals of freedom, justice, and equality to the relational base of the search for human good and basic human morality, it gives these abstract words, for the first time, the emotional resonance that inspires affinity, connection, trust, and pledges of support. This is, in effect, the necessary intertwining of the 'true' and the 'good,' creating both emotional and reasoned support for the ideal of equality. As a result, people's capacity to define new definitions of social change, using critical thought and relational resonance, gives new life to compassion, justice, and democracy. Knowing this, we can begin to organize in new ways, and the culture shifts can begin.

Find Your Group

Everything starts with finding your group. From an effectiveness standpoint, I cannot emphasize enough that this deep participation work, at each level, must be done in groups. It cannot be done alone or individually. You can start reading alone, but you will need to immediately start looking for and finding your group. Find potential members in various places: perhaps a volunteer group; maybe an environmental or justice group; a music group; a boxing group—all are possibilities. Maybe you will find individuals to join a deep participation group in a church, mosque, synagogue, temple group, or book club; or maybe just your favorite wine and whine group, your neighborhood, or circle of friends. The reason and the necessity to explore these questions in a group is that it must be a participatory experience—a *lived* participatory experience.

When beginning your group, it's good to keep in mind both the social and connectivity process. The connecting process is succinctly described by a theory concerning the flow of water and the flow of communications as well. The percolation theory states: "as the number of links in a network gradually increase, a global cluster of connected nodes will suddenly emerge." A 2021 *Scientific American* article, "The Math of Making Connections," explains how this phenomenon "where you really go from local to global connectivity" actually works.[10] At first, it's just a few groups (or nodes), each working separately. But looking at it from the deep participation perspective, as the number of groups slowly increases, differentials diminish, and the flow of *connection* builds. One day it's just a few groups, and six months later, it's everywhere. It's an exciting phenomenon and perhaps explains how culture shifts happen. As word gets around, more people will ask to join your group. Instead, ask them to start another deep participation group, but to stay attached, coordinate, and meet together regularly, perhaps every quarter.

Going forward collectively and beginning to dig into the ramifications of the four ancient wrongs, it will become apparent that this process also requires a *singular fearlessness* on the part of each one of us. To fully understand the social, cultural, political, and economic implications of these ancient wrongs, from both the group and individual perspectives, is difficult and even gut-wrenching work. It means being open to change; open to truly hearing the stories of others; and open to tearing apart some of our most beloved opinions and beliefs. I've said this before. Despite this sometimes-difficult work, I know, through both experience and observation, that this fearlessness is often exhilarating, and the struggles are streaked with joy.

But again, you can't do it alone, or on Facebook, or even at a weekend conference. It requires sustained group work *and* action over an extended period of time. It necessitates inclusion, emotional connection and resonance, shared critical thinking, and collective altruistic action. And yes, these are all things or efforts that we rarely seem to have time for in our busy day-to-day life. But don't be discouraged—it is never exhausting. In fact, it is always replenishing and group-sustaining. The collective *social energy* that emerges from the group work motivates us, connects us, and energizes us. And when we begin to look around with greater awareness, we begin to recognize the increasing numbers of social integrative power displays flashing into action—at least for a moment or two. If we keep practicing deep participation, that momentary flashing action becomes sustained, and change begins to happen.

The Three R's

This is where collective thought and action come together. Working below the decibel of politics, as poet Tracy Smith advised us earlier, deep participation groups can create social integrative power to be used at three different levels—Reform, Reinvention, and Reimaging. The resulting social power adjusts, recasts, and creates new and repaired thought-worlds, awarding new positive thought-worlds the social legitimacy and license to support the operation of the implicitly repaired guiding institutions. Depending on the subject selected by the deep participation group, there are three types of action from which to choose.

Reform actions support proposed change by illustrating the lack of justice in a current policy, law, or practice, then identifying the basic human morality and ethics encased in the proposed change, thereby changing collective understanding.

Reinvention initiates both collective discussion and altruistic action to dismantle negative thought-worlds underlying working institutions. Selected action rebuilds these existing institutions by intertwining the existing functional elements and structures with the new desired aspirational elements

Reimaging brings together profound shared emotion with new ideas to create new concepts, perceptions, and ways of doing. Selected actions utilizing social power can then create desired culture shifts for what was previously believed to be unchangeable (think war or poverty).

Reform is the first method to consider within your deep participation group. While all of the three R's are energizing and many of them just downright fun, reform methods are the easiest to access. While seemingly intensely political, when reform measures are taken out of the political power structure and placed within the perimeter of the social, views change. The moral and ethical human substructure of *voting rights*, for example, becomes immediately apparent. In a country that claims to be democratic, every citizen, as designated in our Constitution, has the absolute freedom and even the duty to vote. It's energizing that at the end of a discussion where the social and basic morality issues are brought in, someone looks at you and says: "Oh, I never thought of it like that."

So yes, it is the basic moral substructure underwriting that deserves consideration, and deep participation Reform initiatives can help citizens step back from the politics of it all to consider the basic human morality underwriting the current policy. The choice is actually quite clarifying. It's a simple yes or no question: Do you believe in equality? In this instance of voting rights, if the answer is 'yes,' democracy and full voting rights are deemed essential for all citizens. If the answer is 'no,' or perhaps 'yes, but,' authoritarian and inegalitarian principles are preeminent, even if not initially recognized. As a result, individual and collective thinking begin to evolve. There are so many other issues, currently deemed political, that can benefit from this stepping back to reconsider the basic human morality of it all—guns, the violence of war, justice, the air we breathe. There are so many issues that can begin to change.

There are other germane reform changes around justice issues to be made which, while more debatable, are still energizing and mind-expanding to consider. Currently, for example, the far-right super majority of the Supreme Court cannot be ignored. Laurence Tribe, the well-known and esteemed Professor Emeritus of Constitutional Law at Harvard University, said as much in a 2021 interview: "Today, The Court appears once again set on an anti-democratic trajectory, one more deeply entrenched and less amenable to a change in course." That anti-democratic trajectory has, in the past several years, only deepened with their recent Dobbs and Immunity decisions. Professor Tribe, with his fears of the Supreme Court's ant-democratic direction, raises critically important questions, ones in which deep participation groups could play an essential role.

Professor Tribe's reflections were based on the findings of the Presidential Commission on the Supreme Court, established in April 2021 by President Biden, of which he was a member. Tribe explains that the general mood of the Commission was one of reluctance, afraid of "being too open about our conviction that a majority of the justices had become dangerously wedded to a political perspective inherently hostile to the premises of a flourishing, inclusive democracy representing all persons equally, and that three members of that majority had been added to the Court by political processes that lacked democratic legitimacy." But the real question was, from Tribe's perspective: should the Presidential Commission deliver a reality-based "slashing critique" or should they opt for the "proverbial noble lie" where the imperfect Court is saved, supposedly protecting the greater truth of our constitutional democracy?[11] The Commission opted for no action whatsoever.

So, if your deep participation group decides that they want to dive deep and reconsider the status of the Supreme Court from an equality and social perspective and enlist other groups, as well, that can make the difference. As Tribe suggests, the one item that would have the most impact is "enlarging the Court," and it would be relatively easy to accomplish. There are no Constitutional rules; in fact, the Court itself has had different number of members at different times. Accomplishment through the required Congressional vote is currently difficult from a political perspective, but working from a social perspective, possibilities change.

Reinvention, the second of the three R's, is fun and enlivening. Once into it, it's all about using shared social energy and practicing collective altruism to reinvent the parts of our communities and society that currently don't work as they should. There are plenty of these non-working parts to choose from, including gun violence, poverty, and an overwhelming emphasis on corporate economic profit. Currently, there are many organizations and voluntary groups doing excellent work in their efforts to eradicate these problems that need to be acknowledged. So, what's so different and enjoyable about social power's reinvention orientation?

The difference for 'reform,' as described earlier, is the change in *perspective* which initiates support for new action and new policies. But for 'reinvention,' the difference is in the *type of action,* and this choice is directly related to praxis as discussed in the previous chapter. Reinvention demands new aspirational types of collective action, often previously considered too idealistic, to be adopted and practiced by the deep participation group, producing both praxis/practice, and a shared social energy. And it is this sustained and shared social energy, inherent in all deep participation activities and actions, which allows aspirational and more idealistic factors to be incorporated into the function of institutions and organizations.

Too often, believing ourselves to be as a human species, primarily individualistic, competitive, and selfish, ideals to the contrary are shrugged off and dismissed. As a result, depending upon behaviors emphasizing generosity, community, and cooperation are thought to be too idealistic, or wishful thinking, and therefore not worth considering. In other words, it is believed that changes of this sort would require the remaking of our highly individualized competitive, selfish, human psyche. But research on economic free-riding vs. generosity among farmers by researchers at Cornell University indicates otherwise. They based their conclusions on two general categories: structural, what is practiced and built-in; and cognitive, what is intellectually recognized, but not practiced.

To begin, the Cornell research validated the idea that practically everyone has inclinations toward both selfish and altruistic behaviors, but the research indicated that in the absence of further influence, most people tend to act according to selfish and individualistic premises. However, the Cornell researchers found the actual situations they *observed* illustrated something unanticipated. Farmers exposed to more generous activities by others turned out differently from what they, as a team, expected. The lead researcher, Dr. Norman Uphoff, explains: "After being ambivalent towards exceptional or exemplary behavior because it was considered aberrant, we came to appreciate it. While encouraging it, we tried to institutionalize it as much as possible into 'normal' (structural) relationships so that what at the time was marginal (or cognitive) would over time raise the average."

OK, I have to stop it here and insert a personal note, I know Norman Uphoff; he and I were professional colleagues for a number of years. And every time I read the above quote I laugh. When Professor Uphoff says he was 'ambivalent' towards exemplary behavior, he really means it. Uphoff's professional perspective has always relied on the quantitative and the absolutely provable. So, in this instance, when he changed his mind and opted for the unexpected because of the evidence, we can be assured that this evidence is iron-clad and gold-standard, so to speak.

So, as the research illustrates, the farmers, with increasing and ongoing association, did opt for the more generous cooperative and social action behaviors, moving these behaviors from the marginal and cognitive to the structural and normal.¹² Another way of understanding this is comparing what the Cornell research group calls the 'marginal, cognitive, and aberrant,' along with my 'aspirational'; compared to their 'normal' 'structural,' and my 'ordinary.' This, in essence, describes where we are today; in the midst of the normal and structural, warily eyeing what, for too long, has been defined as the marginal, cognitive, aberrant, and aspirational.

As a result, it is eminently reasonable that, Uphoff concludes a change in values (e.g., from self-interest norms to altruistic norms) is not the issue. As he tells us, "What matters is not which values one has—we all have many—but which values are activated and applied in a given situation."¹³ Evidently, changing from a set of behaviors emphasizing selfishness—or any other negative practice—to a second set of behaviors emphasizing generosity behaviors (or any other positive practice) feels good, particularly with the supporting social energy of a group. In addition, changing from a set of behaviors emphasizing selfishness—or any other negative practice—to a second set of behaviors emphasizing generosity behavior (or any other positive practice) evidently does not require remaking the human psyche. Instead, it simply requires an expansion of choices within the group, along with an increase of repetition, emphasizing the newly chosen value choice. That we can easily do.

In my experience when beginning to discuss reinvention, many of us imagine quite a strenuous initiative, taking much time and energy, which it can certainly be. But there are other options. One of my favorites that I only read about, is what a group can do, just by sitting and watching. Groups of five to ten people can go to their local courtrooms to observe lawyers and those charged with crimes as they decide whether to go to trial; or whether to take a plea bargain which offers less prison time than if found guilty. Currently, up to 97% of those charged take the plea bargain, often if guilty or not, bringing mounting criticism.

However, there is a slim slice of real-world grounded evidence that illustrates how existing actions and processes can be reinvented. One way in term of more equal justice, for example, is to gather that group of five to ten people together and begin to show up with great regularity at the pre-criminal trial sessions where plea bargains are offered by the prosecutor in front of a judge to the person charged. Evidently, according to several stories, just the sheer observation by groups of people who said nothing and did nothing, other than watch, begins to change the tenor and the outcome of the proceedings for the better. In brief, it seems to me that these group observations might be classified as the beginning of 'Reinvention' as they germinate *change of action*.¹⁴ So, I urge you to be imaginative.

*Reimaging i*s a bit different. In our American culture, we primarily begin problem-solving with the intellect. But reimaging is different, it *always* requires we begin with emotion. In this sense, within our own culture, it is often music that pierces our intellectual view of the world and lets us envision, for a brief moment, a different, peaceful, and bountiful world. This is what reimaging is all about—seeing a path to a different future. If it's about war, sometimes the way to start is to remember, through shared singing and chants, what peace could be like for all of us.

After that, it's easier to move on to the intellect, having used other faculties to reimage the possible. If it's about the environment and our planet, reimaging can start from a place of connected emotion for care and reciprocity. For many of us, the Native American dances that attempt to make peace with the lands begins a different way of relating to the world and life around us. Once re-imaging is moving forward, practicing a particular facet of it through re-invention makes sense as a beginning process. Whatever it is, reimaging is a critical facet to cultivate.

One final point should always be held in mind as we begin deep participation work with the three R's. In this 2020 decade, we are in an extraordinarily charmed and lucky era. I understand that this is difficult to believe, given all of the injustice and suffering that surrounds many of us. But it is true for two reasons. First, we still have, in 2024 at least, a democracy with all of its rights and freedoms, allowing us to practice these three R's. But there is a second reason we cannot allow ourselves to forget—we dare not forget. There is a fourth R, *Resistance,* that is currently rarely necessary to practice. This is because, well before all of us began our activities in this 21st century and earlier, there have always been those stalwart few who practiced Resistance, often under great danger and duress.

It is definitely because of them that we now have the knowledge and awareness which allow us to see the inequality through-line that intertwines the four ancient wrongs. And because we now begin to "see with other eyes" and "hear with other ears," we finally understand how inequality has been supported by our societal acquiescence. Those stalwart few sacrificed much for us to arrive where we are now, within this charmed era, where we can potentially achieve those desired culture shifts. So, we remember what they accomplished while pursuing what we can now accomplish.

Umbrella Leverage

Umbrella leverage and collaborative strategies are key to achieving those necessary culture shifts. Organizing for fearless change and using deep participation's six elements as our preferred methodology will be different from your ordinary politically-oriented protest and social movement. Collective critical thought and emotional resonance create new concepts and new connections. The resulting social energy and shared altruism are then used to bring forth social power that reforms, reinvents, or reimages. But still, like all good plans, deep participation methodologies do require a defined strategy for success. To achieve this maximum impact and success, there are several basic issues that should be taken into account.

The participatory practice always begins with the six elements of deep participation in small groups. Phones can't connect you, nor can Zoom, except in difficult Covid-19-type situations. Deep participation groups, as mentioned earlier, can be formed from sports groups, sodality prayer groups, book clubs—you name it, you invent it, whatever works. The ideal number is seven to eight members, but groups can also effectively range from four to twelve persons. Smaller or larger than that diminishes effectiveness. Heterogeneity of all kinds helps—class, age, sex, race, economic status—but is not required. Making group rules defining honesty, respect, inclusion, and group decision processes, all facilitate the humor, compassion, and laughter which are inherent parts of effective meetings.

However, the most important collective decision your group will make is how to organize under the umbrella of the four ancient wrongs. By doing so, individual efforts of deep participation groups will be multiplied into a larger collaborative movement of strength and impact—and that is the *necessary umbrella leverage*. This leverage remains critical throughout: it maximizes impact and increases future areas of leverage. Equally important, achieving this umbrella strategy for maximum equality impact is relatively easy.

There is a reason that this umbrella strategy works. It is the shared understanding that all of these ancient wrongs, as they manifest in our society and communities, are intertwined and connected. Because the interrelatedness of the four wrongs is basic to the overall patterns and dynamics sustaining the overall structure of inequality in society, it is always necessary to keep that recognition as a basis for group discussions. On the other hand, understanding this overall dynamic does not stop a particular deep participation group from deciding to only choose one, or even a small subsection of one of the ancient wrongs to focus upon. That is perfectly OK because each individualized deep participation group strongly contributes to the overall umbrella effort. Focusing on the ancient wrongs and their intertwining connections of inequality expands maximum impact.

With the individual deep participation group working well, attention can then turn to encouraging new deep participation 'start-ups.' The objective is to develop alliances among new deep participation groups within the community, thereby creating a shared and expanded deep participation social movement base. To start, widely communicate at some selected meeting location what your deep participation group is doing. If, for example, the group has chosen 'peace' as an issue to be addressed, organize an evening meeting at a place that welcomes such public gatherings—libraries, places of worship, schools, or city government buildings—to present your experiences and talk about future planned actions. After the discussion, ask interested attendees to consider organizing their own deep participation groups with the intent of aligning together. They might choose 'cities of peace,' similar to your groups, or something entirely different, such as 'sustainable environments.' But the alignment holds because it is around deep participation, the four ancient wrongs, and the umbrella equality strategy.

Once there are a handful of allied deep participation member groups, discuss and decide to contact more established and traditional community groups—Red Cross, Kiwanis, CASA, Restorative Justice—the list goes on—and arrange to jointly sponsor an annual or semi-annual city-wide weekend conference organized around the injustice of the four ancient wrongs, featuring what *all* the groups are doing around these issues. At the same time, devote part of the actual conference itself for meetings to explain deep participation: how these groups always stay within the social perimeter and the advantages thereof, encouraging interested people to establish their own deep participation groups.

At the same time, always keep the discussion and the possibilities of collaboration with the more traditional and established community groups open and ongoing. Obviously, there are many extraordinary ways to go forward, and your group can add more. Then it begins to take off: to the next community, to the next state .... But never forget to always keep your particular deep participation meeting and your group's collective altruistic action moving deeper and deeper.

Culture Shift Resources

Resources available to deep participation groups are large and diverse and fall into two different categories: external and internal. I name the most basic general resources here, but you will find others that are in your locality or the resources that fit the ancient wrongs your deep participation group has chosen for its focus. Most important, because deep participation is a natural dynamic, it is important to recognize that multiple groups have practiced, and continue to practice, parts of the six elements in culturally appropriate ways—without understanding the entire mechanism. As a result, diverse organizations and initiatives are available to be allied with. For example, long-enduring social movements such as the civil rights movement, the peace movement, or established social-help organizations, all have their roots based on an intuitive understanding of deep participation dynamics. As a result, they represent potential willing allies. And there are many of them, both locally and nationally.

Multiple external resources also exist. Information available in libraries, current books, and internet-assisted research on the issues, are too often easily discounted. For example, as I began this book as one individual researcher,

I was surprised at the depth of information research resources available—including for those of us not known for their digital skills. I was amazed, for example, to find online resources such as "Ask a Librarian" at the Library of Congress where they will actually research specific law and legal questions, and then get back to the questioner with answers. Some states also have similar resources available.

Finally, there is one factor to always keep in mind whether dealing with external or internal resources, and that is the reality of stability vs. instability. Whether we are acting as activist, student, researcher, or observer, we must always be aware of the ground beneath our feet. How stable is it? How unstable is it? The answer makes a difference in how we move forward and what resources we choose to use.

Internal resources are the most important. They begin to make themselves available as the group begins, following the six-element deep participation format. Each of the altruistic-action resources identified below has been discussed in previous chapters. But it is helpful to briefly outline each again, making them easily available for multiple deep participation initiatives. To begin, always remember that the through-line analysis of negative difference and negative indifference examining the ancient wrongs together, provides the two critical transformative guideposts for successful change. They are: I. "*Equality is non-negotiable*"; and II. "*Social relations are primary.*" The following four factors underwrite this culture-shift, transformative agenda.

Initiating Deep Participation Practice. Identifying and discussing the negative thought-worlds which accompany the four ancient wrongs is always the first starting place for deep participation practice. It then helps to consider individual and collective societal assumptions based on these thought-worlds. And never forget the strong social acceptance of *inequality*, which continues to enable the functioning of the four ancient wrongs. As deep participation group discussions begin to focus on how to change the negatives, better collective self-knowledge of ourselves, our community, and our society is slowly built. Remember that opinions are built on sets of assumptions which rightly or wrongly assist in interpreting the facts. Throughout these discussions use the two guideposts which also keeps the focus on necessary exit strategies. Every so often, stop and review these exit strategies. Keep going. Finally, collective meaning begins to appear. But don't stop there. Add in some collective altruistic action to further clarify the idea under discussion. And then keep going.

Altruistic Action. The impact of collective altruistic action on both the givers and the receivers is immense. Words don't do it justice, so the best thing to do is personally experience its transformative factors. Within deep participation, collective altruistic action accomplishes several objectives: generally, it always contributes to the social energy that sustains deep participation over the long term. More specifically, the altruistic action itself begins to clarify what is needed to solve the ancient wrong—or at least how to reinvent part of it. As this practice of clarification emerges, it indicates the ongoing change in the thought-worlds of the deep participation group itself. The way forward in building new and better institutions for the everyday world becomes clear.

Social Power. Social integrative power (also called social power) creates trust, essential for legitimacy. It is only deep participation practice, through expansion of social power, which offers the methodology to collectively establish the necessary sacrosanct societal agreements for a just and caring society. The critical point is that while political and economic power necessarily depend upon trust, they cannot create it. As a result, only social power working within its social institutions provide the necessary "*license to operate*" within society and communities. Neither economic exchange power nor political threat power have this capacity.

Praxis and Practice. Generating collective critical thought through deep participation practice, bringing exhilarating new ideas into the world as a group, fusing them with emotion, and recognizing their importance with a true visceral reaction, is exciting and even individually life-affirming. But it is not enough. We must still make those ideas work collectively for us in everyday life. It's simply not adequate for only the experts, the scientists and philosophers, to define and apply these ideas. We, the people, also need and want to formulate ideas and experiences because it is *we the people*, who can put these new concepts and understandings—sometimes singing to us through the arts, the music, and our practice—into action. Our evolution, and revolution, from Fence-straddlers to wannabe Includers, to Includers, and then doubling back to bring the Excluders with us, is always a revolution of thought and practice.

So, that's it—this is where we are now. Several centuries ago, the politics of democracy created awareness of the inviolability of individual rights. Now there is a second transformative task. Our collective planetary and individual well-being requires that we create a new inviolability—the equality and connection among us all and the Earth itself. In other words, we must transcend; we must make ourselves capable of creating that culture shift.[i] And so, dear reader, practice deep participation, have fun doing so, and watch the world change.

---

i Check in with my website—pauladonnellyroark.com—for my weekly blog offering continuing ideas on how to practice equality and freedom together, allowing us to make those desired culture shifts.

Notes

1 Charles Tilly and Lesley J. Wood, "Chapter 1: Social Movements as Politics", *Social Movements 1768–2008* (Boulder: Paradigm Publishers, 2009), 1–15.
2 Frances Fox Piven and Richard A. Cloward, *Poor People's Movements: Why They Succeed and How They Fail* (New York: Vintage Books, 1979), 36, xiii.
3 James M. Jasper, *The Art of Moral Protest* (Chicago: University of Chicago Press, 1997).
   Burt Klandermans, "Mobilization and Participation", *American Sociological Review,* (Thousand Oaks, CA, 1984) 49, 583–600.
   Dirk Ogema and B. Klandermans, "Why Social Movement Sympathizers Don't Participate" (Semantic Scholar, 1994), 703–722, https://semanticscholar.org
4 Tom Mertes, ed. *A Movement of Movements: Is Another World Really Possible?* (London: Verso, 2004).
5 There are numerous data collections on this subject. I like Wikipedia's "List of Protests and Demonstrations by Size", 2020 because of its specificity in terms of place and subject. It illustrates the diversity and action of social movements.
6 Erica Chenoweth, *Civil Resistance: What Everyone Needs to Know* (New York, Oxford University Press, 2021), 13–14.
7 Robert Savio, "Farewell to the World Social Forum" (Thousand Oaks, CA: Sage Journals, 2019). https://sage.org
8 Will Hobson, "The Concussion Files: Broken promises, Shattered dreams", *The Washington Post,* Feb. 4, 2024, https://washingtonpost.org
9 Anne Applebaum, *Twilight of Democracy: The Seductive Lure of Authoritarianism* (New York: Knopf Doubleday, 2021).
10 Kelsey Houston-Edwards, "The Math of Making Connections", *Scientific American,* Apr. 2021, 24–27.
11 Laurence H. Tribe, "Politicians in Robes", *New York Review of Books,* Mar. 10, 2022, 41.
12 Norman Uphoff, *Learning From Gal Oya: Possibilities for Participatory Development and Post-Newtonian Social Science,* Ithica, New York, Cornell University Press, 1992, pp 326–356.
13 Norman Uphoff, *Learning From Gal Oya,* 337.
14 There are numerous articles on-line about the injustice of this process; so just choose your preferred justice topic. As to the article on group observation, I have not yet found it in my files. But I will, and when I do I will post it on my website.

Bibliography

Applebaum, Anne. *Twilight of Democracy: The Seductive Lure of Authoritarianism.* New York: Knopf-Doubleday, 2021.
Chenoweth, Erica. *Civil Resistance: What Everyone Needs to Know.* New York: Oxford University Press, 2021.
Fox Piven, Frances, and Richard A. Cloward. *Poor People's Movements: Why They Succeed and How They Fail.* New York: Vintage Books, 1979.
Hobson, Will. "The Concussion Files: Broken promises, Shattered dreams", *The Washington Post,* Feb. 4, 2024, https://washingtonpost.org
Houston-Edwards, Kelsey. "The Math of Making Connections", *Scientific American,* Apr. 2021.
Jasper, James M. *The Art of Moral Protest.* Chicago: University of Chicago Press, 1997.

Klandermans, Burt. "Mobilization and Participation", *American Sociological Review*. Thousand Oaks, CA: Sage Publications, 1984.

Mertes, Tom. ed. *A Movement of Movements: Is Another World Really Possible?* London: Verso, 2004.

Ogema, Dirk and B. Klandermans, "Why Social Movement Sympathizers Don't Participate", *Semantic Scholar*, 1994, https://semanticscholar.org

Savio, Robert. "Farewell to the World Social Forum", 2019, https://greattransition.org

Tilly, Charles and Lesley J. Wood, "Chapter 1: Social Movements as Politics", *Social Movements 1768-2008*, Boulder: Paradigm Publishers, 2009.

Tribe, Laurence H. "Politicians in Robes", *New York Review of Books,* Mar. 10, 2022, https://nybooks.org

Uphoff, Norman. *Learning From Gal Oya: Possibilities for Participatory Development and Post-Newtonian Social Science*, Ithica, New York: Cornell University Press, 1992.

INDEX

abortion 40, 42–44, 47, 54
Abrams, Elliot 91
absence of justice 235, 258
abstract words 26; *see also* cerebral patriotism 260
ACLU (American Civil Liberties Union) 23
acquiescence 4, 135, 181, 252, 257; *see also* social acceptance
action-steps 234
Adams, Abigail 39, 64
Adams, John 39, 64, 188
AFDC (Aid to Families with Dependent Children) 106
Africans 17, 60, 71
Age of Complexity 221
Age of Enlightenment 220, 226
ALEC (American Legislative Exchange Council) 126–128
Algeria 77
alt-right 96, 235
altruistic norms 265
AMA (American Medical Association) 118
American Enterprise Institute 110
ancient wrongs 4–12, 28, 31, 113, 123, 142, 147–153, 168, 181, 187, 194–201, 205, 211, 218, 252, 261, 266–269; *see also* four ancient wrongs
Anderson, Carol 25

anger 2; *see also* righteous anger; self-righteous anger
Anti-Defamation League 26, 96
arbitrary power 185, 235
arms sales 78, 90
authoritarianism 7, 93, 189, 198; *see also* violence
Ayatollah Khomeini 90

Bacon, Francis 61, 220
Balibar, Etienne 164, 228, *see also* Equaliberty
barbaric cruelty 185
Barr, William H., 91
Basic Call to Consciousness 139, 141
basic morality 262
behavioral missteps 201
Belew, Katherine 92–96
belief systems 6, 108, 139, 163, 178
Bell, Derrick 35, 98
belonging 5, 53, 92, 163, 173, 181, 209, 224–228, 231, 233, 247; *see also* connections
Boas, Franz 69
Bohm, David 158, 245
Boland Amendment 89, 94
Boulding, Kenneth 5, 160
Breonna Taylor 178
Brown, Michael 27
Buddhist 3, 224
Bush, H.W., 91–93, 131, 189

274 Index

carbon dioxide 124–126, 128, 131, 134, 136
carbon emissions 126, 128, 130–137; *see also* climate crisis
carbon pollution 127; *see also* environmental plunder
Carson, Rachel 129
Carter, Jimmy 130
Castile, Philando 26–27
Catholics 67–69
CDC (Center for Disease Control) 44, 46
Celts 68
cerebral patriotism 260; *see also* abstract words
change agent 11, 147, 159
change of type 199; *see also* participatory change dynamics; readjustment change
Chenoweth, Erica 257
Chief Financial Officers Act 81, 83, 89, 91, 126; *see also* DOD (Department of Defense)
China 81, 125, 134–136
CIA 83
Civil Rights Act 69–70, 149, 190
classism 4, 9, 47, 54, 59, 63, 65, 69, 71, 74, 104, 172, 174, 219; *see also* ancient wrongs
clean energy 128
climate crisis 12, 126, 131, 138, 160; *see also* environmental plunder; Gaia
Clinton 40, 106, 189, 192, 194
coal, oil, and gas 123, 125, 127, 135–137
collective altruism 3, 264; *see also* deep participation's six elements; reciprocal care
collective change of consciousness 1; *see also* culture shifts
collective social energy 239, 245, 250, 261; *see also* deep participation's six elements
collective social learning 248; *see also* deep participation's six elements
collective 1–12, 51, 53, 79, 123, 131, 148–151, 154, 155, 159, 161–165, 169, 199–121, 206, 217, 220, 226, 235, 238, 243, 245–252, 257, 260–264, 267–270; *see also* hyper-individualism
colonization 31, 33, 87, 151, 155, *see also* negative difference

colony 60–163, 72, 151, 184
Columbine 93, 157; *see also* war at home
complicity 52, 92, 206–208, 210
compromise(s) 2, 6, 169, 178, 181, 186–192, 194, 208, 211, 227
conflict 7, 70, 84, 164–166, 168, 199, 218, 226, 231, 245, 251
conflict tradition 154, 164, 240, 244, 248
connections 2, 6, 10, 22, 43, 92, 149, 155, 158, 191, 199, 211, 221, 223, 245, 247, 259–261, 267–270; *see also* belonging
consumption 125, 136, 241
Contra affair 88–90
contradiction 90, 157–160, 172, 198, 221, 227, 229; *see also* stability/instability
COP (U.N.Environment Conference of Parties) 133–135
Copernicus 135, 220, 236; *see also* Age of Enlightenment
corruption 87, 91–93, 127, 153; *see also* complicity
Covid 20, 102. 107, 111, 115, 118, 267
Crenshaw, Kimberle 51–53
critical thinking 6, 161, 246–248, 260; *see also* emotional resonance
culture 1–13, 18, 24, 30, 42, 51, 53, 59, 64, 66, 68, 78, 99, 123, 129, 138–141, 149, 152, 154, 155, 163, 170–172, 177, 187, 193, 195, 197, 200, 211, 218, 233–236, 238, 240–242, 244, 247, 252, 256, 260, 266, 269
culture shift resources 268–270
culture shifts 2, 5, 8–10, 123, 138, 154, 163, 170, 195, 218, 233, 236, 245, 252, 256, 260–262, 266
cynicism 156, 236, 238; *see also* rationality

Daly, Herman 139, 141, 207; *see also* Ecological Economics; environmental crisis
Davis, Angela 30
death penalty 44, 116
deep participation social movements 258, 260; *see also* new social movements; traditional social movements

deep participation 2–8, 12, 119, 142, 147, 149, 154, 163, 199, 218, 221, 233, 239, 242, 244–253, 258, 260–264, 266–270; *see also* research initiatives
deep participation's six elements 245–267
demise 155–160, 164, 198
democratic legitimacy 263; *see also* legitimacy
development 10, 48, 78, 140, 151, 155, 164, 221, 228
discretionary budget 175
dispensation 30–31; *see also* Kendi, Ibram X.
disruption 94, 163
'distinct institutional system', 152, 177, 195, 206; *see also* negative difference
distrust of government 201, 205; *see also* double-talk
diversity 55, 115, 202, 225, 231, 246
Dobbs decision 42; *see also* Lawrence Tribe
DOD (Department of Defense) 79, 81, 86, 99, 175–178; *see also* NDAA
domestic terrorism 96–98
domination 42, 47, 65, 150, 201, 210, 218
double-talk 170, 175–179, 202; *see also* Fence-straddler
Douglas, Mary 2, 161, 249; *see also* social theorists
Dubois, W.E.B. 30
Durkheim, Emile 6, 154, 162, 165, 244, 248; *see also* social theorists
dynamics 3–5, 8, 12, 49, 199, 239, 241, 243–246, 252, 267, 269

Eagleman, David 162, 222
Ecological Economics 139
economic justice 193
economic power 2, 11, 147, 150, 160, 164, 169, 186, 190, 203, 210, 218, 230, 238, 256, 259, 270; *see also* political/economic power
economic recession 1968–1970, 190
Economics Nobel Committee 242
EITC (Earned Income Tax Credit) 106, 110, 112
El Salvador 89
Ellsberg, Daniel 84
Emerson, Ralph Waldo 66–67

emotional resonance 155, 169, 245, 248, 260, 267; *see also* critical thinking
"emotional wallop" 202, 206
empathy 3, 18, 218, 224
endless war 12, 77, 87, 92, 98, 132, 148, 153, 175, 201
enforcement 23, 27, 95, 98, 151, 195; *see also* social enforcement
enlarging the Supreme Court 263
environmental crisis 7, 123, 129, 136, 138–140, 205, 207; *see also* environmental plunder
environmental ethos 138–142
environmental justice 107, 119
environmental plunder 1, 123, 135, 138, 153, 201, 208
Environmental Protection Agency (EPA) 130, 242
Equaliberty 228, 230; *see also* Balibar, Etienne
equality 8, 12, 17–19, 21, 28, 43, 45, 47, 52, 60, 64, 66, 69, 72, 74, 77, 83, 92, 106–109, 111, 119, 138, 141, 149, 152, 169–174, 177, 184, 186–190, 192–195, 197–200, 204–206, 209–212, 217, 219, 221, 225–231, 234–236, 238, 253, 258–260, 262, 269; *see also* inequality
equality based on inequality 198; *see also* paradox syndromes
equality will destroy liberty 172, 189; *see also* Richardson, Heather Cox
equity 137, 203, 212, 227–229
ethics 41, 162, 235, 237, 257, 262
eugenics 67
eviction 117
Excluders 4, 73, 169–171, 175, 177, 184, 190, 194, 198, 200–202, 210, 237, 260, 270; *see also* Fence-straddlers; includers
exclusion 3, 4, 32, 35, 69, 71, 149–153, 160. 171, 191–195, 211, 240, 260
exclusionary template 194
executive power 87–91; *see also* Reagan administration
exit paths 253
Exxon 126–129, 207

Fairness Doctrine 90
false realities 148; *see also* collective social learning; negative thought worlds

fearless change 1, 8–11, 218, 267; *see also* culture shifts
feminism 45, 47, 55
Fence-straddlers 31, 169, 178, 207, 234; *see also* excluders; ideals; includers; wanna-be Includers
forgotten permissions 178; *see also* negative compromise
fossil fuels 123–128, 135, 137
four ancient wrongs 4, 6, 8–12, 28, 31, 113, 147–153, 168, 183, 187, 195, 197–199, 205, 211, 218, 252, 261, 266–269; *see also* ancient wrongs
fractured culture 193; *see also* economic justice;
economic recession 1968–1970
Freddie Gray 27–28; *see also* racism
Freire, Paulo 248
Fulbright, William 87–88

Gaia 134, 138; *see also* Basic Call to Consciousness; *Gaia: A New Look at Life on Earth*; Lovelock, James; Six Nations
Gaia: A New Look at Life on Earth 122, 124
Garner, Eric 27–28; *see also* racism
Garrison, William Lloyd 30
Gazan 91–92
gender 39, 41, 43–49, 51–55, 59, 107, 110, 119, 148, 194, 202, 214; *see also* LGBTQ; sexism
generosity 225, 250, 264
George Floyd 20–23, 25, 157, 163, 224, 257
Ghana 77
Giridharadas, Anand 206, 208
Glissant, Edouard 225, 231
good 234–239, 241, 244, 249
governing philosophy 185; *see also* compromise(s); moderate faction; moderates; moderation
Grassley, Charles 55, 81
Great Society 86
Green energy 125, 128, 137, 207; *see also* solar and wind
greenhouse warming 126; *see also* carbon emissions; carbon pollution
groups (deep participation) 260–268
guideposts (equality and social relations); *see also* resets

Habermas, Jurgen 248; *see also* Freire, Paulo
Hamas 91, 176; *see also* complicity; Gazan
Hansen, James 126; *see also* environmental crisis
hierarchy 48, 63, 152, 158, 169, 177, 201, 203, 209; *see also* 'distinct institutional system'
high differentials 247; *see also* deep participation's six elements
Hill Anita 40–42, 189; *see also* Thomas, Clarence
Ho Chi Min 86; *see also* Vietnamese Freedom Delegation
homelessness 105, 114, 117, 153
Honduras 88; *see also* Contra affair
hooks, bell 41, 45, 47, 218; *see also* love ethic
human psyche 264; *see also* values
hyper-individualism 201, 220

ICE (Immigration and Customs Enforcement) 27
ideals 77, 83, 170, 219, 231, 235, 237, 250, 260, 264; *see also* double-talk; love ethic; mutuality
identity 5, 45, 51, 53, 60, 67, 70, 72–75, 118, 149, 161, 173, 226
imagination 12, 24, 27, 42, 225, 231, 246
Immerwahr, Daniel 79
immorality 64, 92, 184, 187; *see also* basic morality
implications 251, 261; *see also* stability/instability
incarcerated people 114, 115–116
Includers 4, 169–172, 200, 224, 270
inclusion 4, 69–74, 161, 174, 212, 225, 246, 249, 261, 267
indifference 9, 27–135, 64, 118, 123, 135, 138, 151–158, 163, 177–182, 188, 200–203, 211, 217–231, 241, 259, 269
individual rights 220, 270
inequality 8, 11, 20, 25, 43, 55, 67, 71, 73, 103, 106, 108, 113, 118, 148–154, 158, 160, 168, 172–174, 177, 179, 181, 186–195, 197–212, 227, 235, 247, 251, 256, 266, 269
inequality compromise 169, 187–192, 227

inequality practice 150, 152, 174, 198, 201, 208
inequality system 148, 206
injustice 1–4, 8, 12, 18, 27, 31, 33, 39, 41, 46, 52, 59, 104, 148, 150–152, 155, 163, 178, 181, 188, 200, 207, 211, 224, 227, 229, 257–259, 266, 268
instability transitions 156; *see also* stability/instability
institutions 2, 5, 8, 11, 22, 46, 51, 110, 129, 147, 153, 156, 158–163, 198, 205, 229, 238, 242, 248–251, 253, 262, 264, 270; *see also* structure of society; structures
interconnectedness 221; *see also* interdependent; relational; solidarity
interdependent 92, 221
interface 2, 147, 163, 209–211; *see also* trust
international terrorism 97; *see also* Bell, Derrick; social enforcement
investigation 7–10, 40, 49, 54, 91, 97, 148, 150, 168, 172, 178, 192, 199, 211, 248; *see also* four ancient wrongs
invisibility 58, 162; *see also* classism
IPCC (Inter-Governmental Panel on Climate Change) 123, 127, 134, 140
Irish 17, 61, 65, 67–70, 117, 151
Isaac Newton 220; *see also* Age of Enlightenment
Isenberg, Nancy 59, 73, 193
Israel 91, 176

jail population 116; *see also* incarcerated people
Jefferson, Thomas 30, 63, 66, 185
Jim Crow 10, 19, 30, 49–51, 69, 71, 94, 116
Johnson, Lyndon Baines 78, 84, 192
joy 250, 261; *see also* altruistic norms; social energy

Kavanaugh, Brett 40, *see also* sexism
Kendi, Ibram X. 29–31; *see also* dispensation
Kerry, John 90, 134; *see also* climate crisis; Contra affair
King, Martin Luther 18, 28, 114, 191, 218
King Charles II 62
Koch brothers 128
Ku Klux Klan 51

leaderless resistance 94
Lee Atwater 191, 193
left/right division 185
legitimacy 5, 9, 12, 147, 156, 160–162, 165, 184, 201, 209, 218, 230, 233–236, 246–248, 251, 258–260, 262, 270; *see also* social power
Lerner, Gerda 48
less than 7, 34, 174
LGBTQ 4, 43, 47–49, 52–54, 114, 173
liberals 12
liberty and equality for all 171, 184
license to operate 147, 161, 163, 184, 218, 233, 247, 270
living forms of solidarity 199
Locke, John 62
love 5, 53, 116, 140, 161, 218, 233, 248
love ethic 218, 226, 231
Lovelock, James 122, 124, 139
loyalty 5, 161
lynchings 25, 35, 41, 69

Maddow, Rachel 84, 87, 96
Madley, Benjamin 31
maintaining life 222
Manchin, Joe 41, 127
Marine Corps 80
Martin, Trayvon 27–28; *see also* racism up close and crimes of being
mass shooting 157
materialist-reductionist 222
Mather, Cotton 30
Mbembe, Achille 35, 151, 155, 177, 181, 187, 197, 206
McDonald, Laquan 27, 227
McFarlane, Robert C. 91
McNamara, Robert 78, 81, 83–86, 92
McVeigh, Timothy 95
meanness and exploitation 173, 179
methane 128
militarization 95
militias 32, 92, 96
Minneapolis 20, 35
mobility optimism 112
Moderate faction 186, 194
moderates 170, 181, 211
moderation 150, 194, 208
moral protest and participation 257
Morrison, Toni 19, 75
Mudimbe, V. Y. 155; *see also* social theorists

mutation of politics 229
mutual genealogies 224–226
mutuality 5, 160, 225, 248

NASA (National Aeronautics and Space Administration) 126
national interests 80
Native Americans 17, 19, 20, 31–33, 71, 243
natural gas 128
NDAA (National Defense Authorization Act) 81, 175
nefarious terms 180; *see also* wanna-be Includers
negative compromise 187, 189; *see also* complicity
negative difference 34, 53, 64, 108, 123, 151, 158, 170, 174, 181, 188, 195, 197, 201–205, 219, 230, 251, 259, 269
negative indifference 269
negative thought worlds 6–13, 113, 148–160, 171, 177–179, 184, 189–206, 229, 246, 259, 262, 26
negotiation 2, 6, 181, 211
Netanyahu 92
new social movements 257
new social system 159
NFL (National Football League) 259
Nicaragua 88
Nichol, Tyre 23, 95; *see also* racism
Nisbet, Robert 154; *see also* social theorists
Nixon, Richard 190
non-negotiable equality 200, 211
nonviolence 10
non-violent movements 257
Normandy France 184
North America 242
NPR (National Public Radio) 21
numbskull decision 194; *see also* economic justice

Obama Administration 26, 71, 81, 127, 132, 176, 192, 194
OECD (Organization for Economic and Community Deveopment) 102–104
Oklahoma City 93, 95
organizing 256, 259, 267
organizing principle 94, 99, 134, 139, 198, 200, 211, 225, 238, 249, 259

Paine, Thomas 2, 11, 142, 154, 197
Painter, Nell Irvin 59, 64, 74
paradox syndromes 172–177
paramilitary policing 95
Paris Accords 134; *see also* Kerry, John
Paris Climate Conference 134
Parkland 93; *see also* war at home
Parks, Rosa 28
participation practitioner 212, 240, 245; *see also* organizing
participatory change dynamics 243; *see also* change of type
Paul, Rand 35
peace 4, 62, 81, 83–85, 114, 126, 140, 155, 217, 237, 266, 268
peaceable bias 84, 86, 93, 99
perpetrators 2, 33, 227
Piketty, Thomas 73, 168–171, 184, 198, 212
plea bargain 265
Poindexter, John 91
police 21–25, 27, 44, 93–95, 118, 163, 178, 255
political calculus 256; *see also* social calculus
political power 2, 44, 73, 152, 186, 190, 195, 198, 219, 247, 262
political/economic power 2, 147
politics of left and right 1
poor countries 164
Poor People's Campaign 103, 114
Poststructuralists 156
poverty 21–25, 27
practice to theory 238, 240
praxis 251, 238–242
President Clinton 40, 189
Presidential Commission on the Supreme Court 263
Prigogine, Ilya 220; *see also* Age of Complexity
profiteering 1, 62, 83, 204
profound participatory dynamics 243
protest marches 255, 258
Protestant 67
public institutions 205
public interest 80, 90
Putin 91, 96, 140

quadruple gaze 52
quantitative 20, 240–243, 245, 251, 265

race 18, 27, 32, 34, 47, 52, 55, 64–75, 93–96, 98, 150, 193, 202, 219, 227, 229, 256, 267
racism 9, 18–22, 25, 28–35, 42, 47, 52, 54, 59, 64–66, 69–71, 74, 104, 119, 148, 151, 159, 172, 174, 191, 201, 203, 209, 219, 257
racism up close and crimes of being 22–27
Radcliffe-Brown, A.R. 199
Rand, Ayn 205
rape 43, 48–50, 52
rationality 150, 236–238; *see also* cynicism
readjustment change 199, 245; *see also* change of type
Reagan, Ronald 131, 190, 193, 205; *see also* Ayatollah Khomeini; Contra affair; executive power
Reagan administration 89
real revolution 2, 4, 11
reciprocal care 108, 153, 206, 228, 230, 253, 259; *see also* belonging
reconstruction 69
reform 200, 246, 262–264; *see also* organizing
Reich, Robert 212
reimaging 200, 262, 265; *see also* organizing; organizing principle
reinvention 160, 200, 246, 262–266; *see also* organizing; organizing principle
relational 92, 221–227, 230, 239, 260
relational mind 221, 224, 226; *see also* Siegel, Daniel
religious morality 235
research initiatives 239, 242, 244
reservoirs of life 197
resets 200, 211; *see also* guideposts
resistance 23, 45, 65, 80, 94, 154, 224, 226, 257, 266
resonating emotion 6, 246; *see also* critical thinking
revolution 2, 4, 11, 63, 71, 93, 94, 123, 142, 154, 185, 220, 222, 231, 270
revolution of thought 270
rhizome identity 226; *see also* Glissant; hyper-individualism
Rice, Tamir 27, 224
Richardson, Heather Cox 172, 177
right to command 152, 177, 182
right to vote 72, 187
righteous anger 3, 200; *see also* Buddhist; self-righteous anger

Roberts, John 41
Roof, Dylan 95–96; *see also* sanctification: guns, war, and violence; war at home
Roosevelt, Franklin 90
root causes 4, 6, 11, 200
Royal African Company 62
Ruby Ridge 92, 95; *see also* war at home

sacred 124, 139, 155; *see also* Gaia; Six Nations
Said, Edward 152, 155; *see also* social theorists
sanctification: guns, war, and violence 96–99
Sandel, Michael 212
Sandy Hook 93
Saul, John Ralston 235
Scientific Revolution 220, 222
Scott, Walter 27
security 18, 82, 88, 91, 94, 97, 115, 131, 136, 228
see with other eyes 2, 197, 231, 266
self-interest 51, 133, 170, 177, 181, 187, 201–205, 219, 221, 265
self-organization 221
self-righteous anger 3; *see also* righteous anger
sexism 41–42, 46–47, 52, 53–55, 172, 218–219
sex-trafficking 49
Siegel, Daniel 223
SIPRI (Stockholm International Peace Research Institute) 81; *see also* arms sales; NDAA
Six Nations 139; *see also* Basic Call to Consciousness; Gaia
Slow progress 188
Smith, Tracy 217, 223, 253, 262
SNAP (Supplemental Nutrition Assistance Program) 134
social acceptance 195, 197, 200, 206, 210, 235, 269; *see also* injustice; social change
social action 2, 8, 10, 154, 265
social calculus 256; *see also* culture shifts; social power
social change 5, 28, 154, 156, 159, 169, 221, 234, 239, 243, 251, 260
social energy 4, 154, 161, 200, 225, 239, 245, 247, 249, 252, 261, 264, 267, 270

280 Index

social energy and altruistic action 249
social enforcement 98
social institutions 2, 159, 163, 270
social integrative power 2–6, 8, 12, 123, 147, 160, 163, 169–171, 211, 218, 233, 239, 245–247, 250, 253, 261, 270
social movements 256, 260, 268; *see also* deep participation social movements; new social movements
social order 12, 147, 160, 195, 201, 253
social power 2–4, 6, 10–13, 123, 138, 141, 147
social relations 5, 46, 156, 157, 199, 211, 230, 237, 258, 259, 269
social self-knowledge of society 11, 183, 246, 269
social theorists 2, 154, 164
social theory 148, 155, 166, 239, 244
social transformation 5, 164, 218, 233, 246
solar and wind 129, 137
solidarity 2–4, 8–10, 45, 53, 55, 154, 161, 165, 168, 199, 206, 121, 212, 219, 226, 240, 244, 246, 248, 256
solidarity base 199
solidarity tradition 154, 165, 240
sprint methodology 187
stability/instability 6, 8, 13, 20, 147, 54, 117, 147, 154–165, 198–201, 221, 228, 234, 239, 246, 251–253, 269
stalwart few 217, 266
Stanford 32, 33, 104, 106–108, 113, 126
starting point 2, 148, 156, 164, 227, 242, 244, 245, 258
start-ups 267
status quo 244, 253
Steady-State Economy 139–141
Stiglitz, Joseph 161, 212
structure of society 172
structures 2, 4, 8, 11, 22, 28, 31, 46, 49, 51, 53, 85, 153, 160–162, 174, 178, 201, 203, 220, 230, 236, 238, 248, 251, 262; *see also* institutions
subconscious thought-worlds 55, 179, 202, 237
Sudan 92
Supreme Court 40–43, 116, 131, 189, 263

SWAT teams 24, 95
systems 2, 6, 21, 39, 42, 46, 48, 51–53, 65, 67, 81, 108, 139, 141, 160, 163, 176, 178, 194, 201, 203, 206, 208, 121, 219, 226, 237–239, 247, 251

TANF (Temporary Assistance for Needy Families) 106, 110, 115; *see also* SNAP (Supplemental Nutrition Assistance Program)
temperature rise 124, 130, 134, 141; *see also* carbon emissions; carbon pollution
Thomas, Clarence 40–42, 189
thought worlds 65, 162
through-line 9, 28, 148–150, 158, 168, 172, 266, 269
through-line of inequality 148, 150; *see also* through-line
Till, Emmit 26–28, 35; *see also* racism up close and crimes of being
traditional social movements 256; *see also* social movements
transformative agenda 269
Tribe, Laurence 263
true change 1, 22, 217
Trump 7, 22, 26, 29, 73, 91–93, 96, 99, 192–194, 205, 209, 256
trust 2, 5, 8, 12, 41, 84, 88, 92, 147, 158, 160, 163, 166, 169, 198, 200, 204, 209, 222, 233, 239, 246, 248–252, 258, 260, 270
truth 156, 158, 162, 165, 170–174, 184, 201, 205, 226, 230, 236, 238–240, 245, 260–263
Turgot, Jacques 220

Ukraine 92, 129, 140, 176
umbrella leverage 266
UNEP (United Nations Environmental Program) 238
unexpected results 243, 245
UNFCC (United Nations Framwork Convention on Climate Change) 133
UNICEF (United Nations Children's Fund) 238
Union of Concerned Scientists 126
United Nations 33, 44, 92, 102, 133, 139, 237
Universalism 173
unstable times 5, 9, 162

Uphoff, Norman 264
utopistic 159; *see also* Wallerstein, Immanuel
Uvalde 92, 157, 224; *see also* war at home

vaguely right 235, 238, 260
values 2, 74, 78, 109, 149, 153, 161, 173, 179, 181, 201, 207, 234, 265
Vietnam War 78, 84, 87, 157
Vietnamese Freedom Delegation 85
violence 1, 4, 7, 12, 25, 32, 43–46, 50, 67, 74, 78, 93–99, 114, 118, 149–151, 160, 168, 173, 179, 198, 201, 209–211, 219, 226–228, 231, 256, 263
voting rights 28, 70, 115, 118, 189, 192, 262
Voting Rights Act 28, 70

Waco 92, 95; *see also* war at home
Wallerstein, Immanuel 159, 173, 177
wanna-be Includers 171, 180, 189, 234
war at home 92–96; *see also* leaderless resistance
war on poverty 106
waste people 60, 66, 71–73, 151, 231
Watt, James 131
We the people 177, 270
West Africa 239, 242, 251
where we are 12, 18, 22, 49, 103, 148, 192, 265 270
white supremacy 51, 96, 99, 169; *see also* racism; white trash; whiteness
white trash 59, 74, 193
whiteness 19, 22, 25, 31, 64–67, 69, 74
Wideman, John Edgar 26, 34
Williams, Caroline Randall 49, 51
Williams, Thomas Chatterton 26
Wilson, Woodrow 85
World Climate Conference 131
World Social Forum 258
world view 7, 10, 12, 148, 184, 194, 211, 218, 219, 220–221, 225, 227

zero-sum games 202

For Product Safety Concerns and Information please contact our
EU representative GPSR@taylorandfrancis.com Taylor & Francis
Verlag GmbH, Kaufingerstraße 24, 80331 München, Germany